VERBAL DECISION ANALYSIS FOR UNSTRUCTURED PROBLEMS

T0189531

THEORY AND DECISION LIBRARY

General Editors: W. Leinfellner (*Vienna*) and G. Eberlein (*Munich*)

Series A: Philosophy and Methodology of the Social Sciences

Series B: Mathematical and Statistical Methods

Series C: Game Theory, Mathematical Programming and Operations Research

Series D: System Theory, Knowledge Engineering and Problem Solving

SERIES C: GAME THEORY, MATHEMATICAL PROGRAMMING AND OPERATIONS RESEARCH

VOLUME 17

Scope: Particular attention is paid in this series to game theory and operations research, their formal aspects and their applications to economic, political and social sciences as well as to socio-biology. It will encourage high standards in the application of game-theoretical methods to individual and social decision making.

The titles published in this series are listed at the end of this volume.

VERBAL DECISION ANALYSIS FOR UNSTRUCTURED PROBLEMS

by

OLEG I. LARICHEV

and

HELEN M. MOSHKOVICH

Institute for Systems Analysis,
Russian Academy of Science

KLUWER ACADEMIC PUBLISHERS

BOSTON / DORDRECHT / LONDON

A C.I.P. Catalogue record for this book is available from the Library of Congress.

ISBN 978-1-4419-4777-2

Published by Kluwer Academic Publishers,
P.O. Box 17, 3300 AA Dordrecht, The Netherlands.

Sold and distributed in the U.S.A. and Canada
by Kluwer Academic Publishers,
101 Philip Drive, Norwell, MA 02061, U.S.A.

In all other countries, sold and distributed
by Kluwer Academic Publishers Group,
P.O. Box 322, 3300 AH Dordrecht, The Netherlands.

Printed on acid-free paper

Printed in the Netherlands

Preface

A central problem of prescriptive decision making is the mismatch between the elegant formal models of decision theory and the less elegant, informal thinking of decision makers, especially when dealing with ill-structured situations. This problem has been a central concern of the authors and their colleagues over the past two decades. They have wisely (to my mind) realized that any viable solution must be informed by a deep understanding of both the structural properties of alternative formalisms and the cognitive demands that they impose on decision makers. Considering the two in parallel reduces the risk of forcing decision makers to say things and endorse models that they do not really understand. It opens the door for creative solutions, incorporating insights from both decision theory and cognitive psychology. It is this opportunity that the authors have so ably exploited in this important book.

Under the pressures of an interview situation, people will often answer a question that is put to them. Thus, they may be willing to provide a decision consultant with probability and utility assessments for all manner of things. However, if they do not fully understand the implications of what they are saying and the use to which it will be put, then they cannot maintain cognitive mastery of the decision models intended to represent their beliefs and interests. If they recognize their loss of control, then they may withdraw from the process and revert to relying on intuition. If they fail to recognize it, then they may accept recommendations whose implications they cannot really follow.

Recognizing this quandary, the authors ground their proposal in a detailed cognitive task analysis of what people need to do in order to translate the images of decisions in their heads into verbal or quantitative terms. The analysis proceeds to review a wide range of cognitive studies, examining people's ability to perform these tasks. The resultant picture is a mosaic of strengths and weaknesses. People do some things very well, other things poorly. What they do well often reflects reliance on heuristics, with a range of applicability. As a result, using them requires either recognizing these bounds or recognizing the fallibility of the resulting decisions.

Within the constraints identified by this analysis, the authors propose an approach that integrates modeling procedures with elicitation methods. In the spirit of the late Clyde Coombs's Theory of Data (1964), they get as much information from people as they are capable of giving. What people say is then used in a modeling framework designed to extract as definitive recommendations as the data will allow - while providing a clear picture of its own definitiveness. They demonstrate the approach in various contexts, including a landmark application to research-and-development planning, a topic of increasing importance in this country. All in all, this volume is ambitious and challenging, simultaneously addressing theoretical and applied problems, in a way that advances each by consideration of the other.

Baruch Fischhoff
Carnegie Mellon University

Table of Contents

INTRODUCTION

What is common in different professions such as state officer, engineer, geologists, physician, or manager? They all involve decision making, that is, choosing one variant of action among several options. These problems are inherent to all kinds of human activity.

This book tells about the exciting world of human decisions and methods of action that improve the chances of making sensible decisions in a complicated, contradictory, and incompletely definite environment. We call a human decision sensible if in a difficult situation the decision makers take into consideration all appropriate factors and plausible outcomes and get estimates of the best experts. In short, all information available at the time of decision making is used by the decision makers from the standpoint of their policies, preferences, and intuition.

Some conclusions can be drawn from this definition. First, the sensible decision is not necessarily unerring from the point of view of impartial future observer. The decision maker (DM) can evaluate incorrectly the course of future events; new circumstances and factors affecting the events can appear already after making the decision, and so on.

Second, sensible decisions in complicated situations are always subjective. The DMs compensate by their intuition the lack of information and by their forecasts the lack of reliably calculated future trends. Additionally, the DMs are often spurred by the deadlines.

Third, a sensible decision must be based on the DM's policy, that is, a well thought-out course of actions, which also can prove to be unsuccessful in view of the future observer who will know the situation and results of decisions made. Nevertheless, if the decisions are to be sensible actions rather than hasty, sometimes useless, jerks, the DMs must have definite policies and preferences. They are not working in seclusion, but are surrounded by their colleagues and collaborators to whom they must explain somehow the logic of their actions. We only note that to their adversaries this explanation can look illogical.

Fourth, we do not regard the word "sensible" as a synonym of "stand-off" or "impartial." On the contrary, a sensible decision reflects (and must do it) the main DM's objectives, no matter how emotional they are. For instance, reasonable politicians must take into consideration the effect of a decision not only on the actions, but also on the feelings of their allies and adversaries. Different subjective and objective factors must make a reasonable combination - as if a combination of the ice of reality and the flame of expectations.

It is, thus, obvious that making sensible decisions is for the DMs an extremely difficult task. "There must be a lucky break in this business!" could say the reader. Yes, but it is our belief that there must be something more than just a lucky break. Methodical analysis of all related circumstances, circumspect examination of all information available, and, finally, special methods for reviewing options of decisions improve dramatically the chances of successful decision making. The aforementioned methods are designed to enable the DMs to understand not only the situation, but themselves, to work out a policy, and to find a tradeoff between the desirable and the possible.

Numerous studies in diverse scientific fields, mostly economics, are devoted to the problems of decision making. In that of the theory of decision making, two major lines of study can be identified:

(a) normative methods of decision making which prescribe the DMs how to make the best decisions and

(b) descriptive studies showing how in real life people act when making a decision.

Below, we will dwell in detail on the results of these studies and seek to demonstrate that the results of both lines of study are at variance with each other. Stated differently, if people are such as depicted by the results of descriptive studies, then it is unlikely that the majority of the existing normative methods could profit them.

This contradiction manifests itself variously in different classes of decision-making problems. There exist problems whose objective content largely defines the best approach to their solution. In this book we address another range of problems where the model is the subjective human perception of the environment.

Our studies proceed from the contradiction between the normative methods and descriptive results. To our opinion, a new approach is required in the domain of decision making; new and sufficiently strict foundation of the decision making methods is required. Since this domain of studies is multidisciplinary, the procedural principles for designing these methods must be multidisciplinary as well.

This book aims to construct anew the procedure of decision making in nonstructured problems and demonstrate some methods that are logical inferences from it. These methods will be shown below to be psychologically valid, because they are based on the results of studying the human information processing system .

We will consider problems of decision making where the main factors are purely qualitative and have no natural numerical equivalent. One of the classifications (Newell, Simon, 1957) calls them nonstructured problems. They encompass political, personal, and some business decisions. As will be shown in what follows, an adequate representation of the content of these problems is extremely sensitive to the methods of describing them - it requires that the natural verbal description be retained at all stages of analysis.

We do not claim that this class comprises all problems of decision making; this is not the case, and we will try to outline the range of problems to which this book relates. We also will show that there is an opportunity of constructing methods of decision making through the utilization only of qualitative (not quantitative!) variables and logic (not quantitative!) operations of data handling.

Therefore, the new class of methods for decision making will rely on an absolutely different procedural basis.

In this book we discuss mostly the problems of individual choice, but this does not imply that here the DMs are dictators doing whatever comes to their heads. Even in their personal decisions people take into consideration the opinions of their neighbors. In sensible business decisions, the DMs consider the opinions of their environment and assistants, analyze the decisions from the standpoint of their acceptability to different groups that could be affected by the decisions. The DMs must make their choice with due regard for all circumstances - that is why the problems of individual choice occur so often in practice.

If we turn to the practical use of these methods, then we should be aware of the fact that there must be a demand for best variants of decisions and critical analysis of the problems of strategic choice. Various factors such as competition or political opposition compel the DMs to try hard in seeking the best variants of decisions and excel in finding the best way out of the existing situation. Otherwise, the demand for carefully prepared decisions and scientifically well-founded methods can be very limited.

This problem has profound personal aspects. The words of Thornton Widler that the ability to choose is the most precious property of the reason come pat to the occasion. It is not an inherited instinct, but the evidence of a free mind which people must develop in themselves. Little can be achieved in this life without permanent exercise, and although personal qualities such as readiness to risk, quest for diversity, etc., are conducive to the ability to choose, they are not sufficient. The society must pose the problems of choice before its members from the very childhood, enabling them to make a free choice of their way and to bear responsibility for its consequences.

The need to choose - the majors for students and jobs, books, friends, or hobbies for adults - is a part and parcel of life. It exercises people and subtilizes their mind and ability to perform a specific human activity, that of decision making.

The general culture of a society and its ability to overcome crises in its development grow with culture of decision making. A deep conviction in social usefulness of the methods of analyzing variants of decisions encouraged the authors in developing the methods and solving practical problems.

Anybody who met in practice with the problems of decision making knows how difficult it is to pierce into their heart, what a truly "detective" work must be done to divide the "*sheep from the goats*" (Math. 25:32) and dissipate the haze of indefiniteness. It is only in the textbooks that the problems of decision making are presented in their "pure" form to which you just have to apply one or another method of solution. In real life, people analyzing problems often meet with difficulties about which little is said in scientific publications, but which are characteristic of almost any problem of decision making.

The real problems of decision making feature a verbal descriptive language, a flexible and reliable tool reflecting the complexity of the world that surrounds us. The natural desire of DMs to interfere with analysis and check all inferences made from their preferences creates the need for "*transparent*" methods and for retaining the qualitative verbal language of description on all stages of problem analysis.

One of the reasons for developing the methods of verbal analysis of decisions stems from the experience gained in solving nonstructured problems. The approach to verbal analysis of decisions as developed under this cover is oriented to studying the nonstructured problem by qualitative logic means, rather than quantitative ones. This book describes new methods of decision making that do away with numbers, and this approach can be expected to seem unusual to many readers.

Indeed, our education is largely number-oriented. "One" and "two" are among the first words that a baby learns to say. At primary school mathematics is taught hand in hand with language, but at secondary school together with physics it forces out literature and history. At universities, the cult of number is dominating. Only history and art sciences avoided it, but only to some extent - it suffices to mention the art measurement and computer simulation of historical events.

Many studies of decision making are based on an implicit assumption, which is shared by many, that human thinking is like that of a true-born statistician. It is assumed that human judgment of a whole must be based on the estimates of its samples. People also must evaluate in numbers the outputs of decisions. These assumptions obviously stem from the desire to evaluate the efficiency of any action.

Yet, not anything can be evaluated numerically even in the world of private entrepreneurship. As we move from the numerical empire of natural sciences to the world of human behavior, we meet with many unmeasurable factors. Love for the neighbor, aesthetic pleasure of contemplating a beautiful painting, or hope for better future cannot be measured in numbers like length or temperature. Such factors as client's reliability, company's prestige, leader's appeal, and many other are qualitative by nature.

People are used to the world where qualitative factors keep themselves in the foreground. Attempts to introduce numbers into this world and substitute the quantitative for the qualitative require another way of thinking. There is another approach which consists in developing methods of decision making that are suited to human language, any qualitative factors, and human procedures of decision making, which is what we tried to reflect in our book.

Although the working conditions of consultants on decision making depend substantially on the situation in one or another country, the methods of decision making are universal. No science can be divided into "national schools." According to Anton Chekhov, "*There is no national science, as well as there is no national multiplication table.*" When working on this book we felt as members of a unique international scientific team.

Acknowledgment

Scientific contacts with members of the International Society of Multicriterial Decision Making, European Group of Decision Making, European Association of Decision Making, and International Society "Judgment and Decision Making" were especially important for us.

This book is the fruit of many years of research, and its authors would like to express their gratitude to those who rendered them assistance over all these years.

Our thanks are due to the member of Russian Academy of Sciences Prof. S. V. Emel'yanov who during many years supported our studies and practical activity.

The making and development of our studies in Russia are greatly indebted to the permanent contacts with Profs. Paul Slovic, Baruch Fischhoff ,Bernard Roy and Patrick Humphreys whose concern and valuable advice were very helpful.

The arthors want to express the gratitude to Prof.D.Olson for many useful discussions on the topics presented in the book,for reading and commenting on some portions of the presented material.

Our colleagues from the Department of Decision Making of the Institute of System Analysis of the Russian Academy of Sciences participated in the development and practical application of the methods of decision making. Our thanks are due to M. Yu. Sternin, A. B. Petrovsky, A.V.Kortnev, A. I. Mechitov, L. S. Gnedenko, E. M. Furems, G. I. Shepelev, and V. S. Boichenko.

The first author is thankful to Prof. B. Roy (France) under whose leadership he made, in collaboration with P. Bertier, his first acquaintance with the enchanting world of decision making.

The first author is indebted to K. Borcherding, F. Klee, and J. Rose for assistance rendered to him at an important moment of his life.

The second author want to express her gratitude to Prof.W.Stein for many useful discussions ,reading and commenting on some portions of the presented material and to Dr.E.Roach for his support for the project and encouragement in her endeavors.

We are thankful to our colleagues B. Fischhoff, A. Vari, R. Brown, and W. Edwards for the comments and advice on the prospectus of the book.

Only the authors bear responsibility for all results presented in this book.

1 INDIVIDUAL DECISION MAKING: APPROACHES AND METHODS

Decision making is a special kind of human activity aimed at attaining the posed objective. If we turn to the history of humankind, we can see that decision making was always an important aspect of human activity.

Before the appearance of first tribes, people had to decide how to choose a cave, where to find food, how to defend themselves against beasts of prey, and so forth. The coming of state gave rise to making administrative decisions. History provides us with examples of great battles and decisions made by famous generals of all epochs and peoples. Over the last three centuries, it was the economists who tried to gain insight into the rules of decision making. In all these decisions, both successful and unsuccessful, the contradictory human nature - intuition and rational calculations, emotional estimation of events and logic thinking - is reflected as in a mirror.

In the human world of decision making, emotions and reason are inseparable duumviri. In personal decisions where the consequences tell primarily on the destiny of persons themselves, emotions often dominate. Quite another matter is making state, legal, military, business, etc., decisions where the decision maker (DM) must be able to explain the reason behind one or another decision. This person must be able to appeal to common values and senses, to religion, logic, anticipated consequences, and so forth. Here, the logic of human decisions comes to the foreground. When explaining their decisions to the others, people can refer to any factors - legal, monetary, or purely emotional; what counts is the consistency of logic inferences from the premises and the convincingness, it is important to be understood and be consistent.

How can we characterize the decision making methods used previously by the most reasonable persons?

They primarily tried to make use of their previous experience, to understand the problem, to obtain all necessary information, and to take into consideration all important factors and reject the collateral ones. This ability in different times was known under different names, but "intelligent common sense" seems to be the most apposite name. To make decisions, people traditionally used problem descriptions in natural languages understandable to their neighbors (when analyzing them) and logic (when solving them).

The XXth century brought into the human history special difficulties of decision making. The progress of science and technology amplified the impact of decisions of different people in different places of the world. Along with interrelations between decisions, objective complexity of each individual decision has grown. Beside the familiar business criteria such as expenses and profits, new criteria appeared such as environmental impact, personnel training, world competition, etc. Extremely complicated installations such as nuclear power plants, chemical plants, or gas pipelines emerged which can bring countless adversities. New problems of managing large institutions characterized by vast personnel and complicated production appeared.

The human practice responded to these challenges by developing decision making methods. Although the problems of decision making are as old as the world itself, their methodical study started only in this century, and only within the last two or three decades it became obvious that these problems are intrinsically multidisciplinary.

When establishing the norms of human behavior in various activity domains, one must be aware of the abilities of man as a biological being having an original data processing system. This system influences very much the human behavior in decision making. Therefore, the economists, engineers, sociologists, and psychologists investigating the problems of decision making must take into consideration the results many scientific disciplines and not only of their own domain.

It might be well to note that historically the studies of human decision making were carried out in different scientific disciplines independently. And even now the results obtained in one domain of science are known insufficiently, if at all, to the researchers working in another domain. Yet, the results of these studies lay foundation for future synthesis of findings and achievements of various lines of science. When combined, they will generate a new quality like arrows aimed to one target or flows of water running from the mountain slopes to one lake. Even today the answers to many questions must be sought in an adjacent domain of study. This need will be even more acute in future.

The conviction that decision making can be usefully regarded as a multidisciplinary domain stimulated us to combine under the same cover several overviews associated with the key word "decision making." Each of the five domains of study discussed in what follows is large and deserves a special study. What we tried to do was to compile a concise Baedeker of these worlds focusing on what is most important in our view.

Notably, this study is confined to the problems of individual choice. We deliberately avoided the problems of collective choice which make up a large independent domain. We believe that namely the problems of individual choice are of prime importance from the point of view of human practice. An outstanding successful American manager L. Yakok emphasizes that despite the textbooks, the most important decisions in corporations are made by individuals, rather than collective bodies or committees (Yakok, 1990). Much can be found in the literature about the style and methods of individual decision making by large state and military personalities.

It is understood that individual decisions can be successful if the manager is not an autocrat but has to explain and prove the decisions. Nevertheless, the choice must be always personified, and the person must be responsible for it. It is namely the desire (and possibility) to disguise individual decisions as those of organizations, committees, or boards and, thus, avoid responsibility that accounts for many unsuccessful decisions.

In what follows, we will display the diversity of problem formulations from mathematics, economics, computer science, and psychology, but first we dwell on the terminology.

Participants of the process of decision making

In weakly structured problems of decision making, the problem of choice proper is closely related to its owner (Vari, Vecsenyi, 1984). The owner of problem is a man who, in opinion of his/her associates, must make decisions and take their consequences. These decisions can take effect on his/her welfare and standing. Yet, the problem owner is far from always being a decision maker (DM), though this can be the case and there are many examples of combining both roles.

There occur situations where the problem owner is only one among the persons participating in its solution.

The case is possible when the DM and problem owner are different people. We all know families where the formal head of the family decides nothing. Similarly, executives often try to shift their responsibility to the others: presidents of companies rely on their deputies, and presidents of states sign, sometimes contradictory, decrees prepared by their subordinates. Thus, the problem owner and DM can be both one or different persons.

Decision making in some way is affected by active groups. By this term are meant groups of persons having common interest in the problem at hand. For instance, if a decision must be made about constructing a nuclear power plant, then the active groups are as follows:

- staff of the Ministry of Power Industry interested in
  increasing production of electrical power,
- staff of the agency responsible for the construction
  of the nuclear power plant,
- representatives of local residents, and
- representatives of the defenders of environment.

In this case, the local authority which granted the building permission for nuclear power plant is the owner of problem and, sometimes, the DM. We know from experience that it is not always the case, which causes many conflicts. A DM of good sense, when estimating alternatives, always considers the interests of active groups, takes into account their positions and criteria.

Experts, that is, professionals familiarized with individual aspects of the problem under consideration better than the DM, play an important role in decision making. They are asked to give estimates or forecast the outcomes of one or another decision. The experts express their subjective opinions; yet, provided they are impartial and versed in their business, their estimates are close to objective estimates.

In case of building a nuclear power plant, for example, experts in physics can provide a valuable information about the impact of the plant on residents and environment. They can evaluate the probabilities of accidents and their possible consequences. It must be remembered, however, that the decision is made by the DM and that the experts only provide a portion of the required information.

To help in preparing complicated (usually, strategic) decisions, analysts of decision making are usually invited. Their role reduces to organizing a rational process of decision making: to helping the DM and problem owner in formulating correctly the problem, to identifying the roles and positions of active groups, and to organizing interaction with experts. The analysts usually do not give their own estimates, but help the others to understand the preferences, to weigh all pros and cons, and to work out a reasonable compromise.

Thus, the persons involved in problem solving can play one of the following roles:

(1) problem owner,
(2) DM,
(3) representative of active group,
(4) expert, or
(5) analyst.

Additionally, the DM's environment, that is, the staff of the organization for which he/she makes decision, participates implicitly in decision making. And it is primarily to this group that the DM explains consistency, rationality, and validity of the decision. It must be mentioned in this connection that, although the DM makes individual decisions, they are influenced by the policy and preferences of this group. This applies to all situations mentioned above. As was noted, a reasonable DM cannot fail to take into consideration the opinions and positions of active groups. The individual nature of decisions that are made with regard for collective preferences is characteristic not only of the two cases where the persons of DM and problem owner coincide completely or partially. Even in the case of a collective organ making important decisions, we observe the desire of its participants to explain their positions, to work out a common policy, to find a common decision. Then, such an organ plays the role of DM having a definite policy.

Alternatives

By alternatives are meant variants of decisions. Alternatives are part and parcel of the decision making. Therefore, at least two alternatives are required for a problem of decision making to exist. Alternatives can be dependent and independent. By independent are meant alternatives over which any action (removal from consideration, identification as the best alternative, etc.) can be executed without influencing the quality of other alternatives. For dependent alternatives, decisions about some of them exert effect on the quality of others. Different types of dependence of alternatives can be identified.

The direct group dependence is the simplest and most natural dependence: if it is decided to consider at least one alternative from a group, then the entire group must be considered. For example, in planning urban development (Larichev et al., 1990) one conceptual possibility (for instance, the transfer of an airfield) requires that all variants of realizing it be considered.

Another type of dependence of alternatives is the dependence on disregarded alternatives. This type is mentioned in the well-known book (Luce and Raiffa, 1957) where an example is presented of the dependence of chosen dishes on those included in or excluded from the restaurant menu. In the general theory of choice (Aizerman, Aleskerov, 1990), the condition of choice independence of the contraction of the set of alternatives was baptized heritage condition.

The third type of dependence is called the dependence on not existing "phantom" alternatives (see, Farquhar, Dratkawic, 1991). The image of an ideal alternative created by a person during choice can influence the choice from real alternatives, especially if there is hope to realize the ideal variant. (We recall the well-known problem of choosing the best secretary in the methodology of successive choice [5]).

Methods of decision making must be adjusted to types of dependencies between alternatives. Analysis of the dependence itself must be a stage of one or another method of decision making.

The problems of decision making can differ substantially in the number of alternatives and in their availability at the time of working out a policy or making a decision.

There exist problems where all alternatives are given and one has only to choose from this set. For example, we can seek a rule for choosing the best items from existing ones, the most efficient organization, best university, most democratic country, best of the constructed yachts, and so on. These problems feature closed nonexpansive set of alternatives.

Yet, there exist a lot of different problems where all alternatives or a part of them are not yet generated by the time of decision making. For example, if it is required to work out a rule for opening of credit line to organizations or private persons, then alternatives (particular organizations or private persons) can appear only after working out and publishing the rules. Similarly, one needs not know all R and D proposals which will come to the fund in order to have a well-formulated policy for their estimation and screening.

When there are many (hundreds and thousands) alternatives, the DM cannot concentrate on each of them, and a need occurs here for

- clear rules of choice,
- experts, and
- facilities allowing one to realize a consistent policy.

All this is required - though not so urgently - also if the number of alternatives is comparatively small (up to twenty). In problems such as choosing a plan of political campaign, or gas pipeline route, or plan of urban development, or fashion style, etc., there are not much alternatives from which choice starts, but they are not the only possible ones. It often happens that they give rise to new alternatives or to a set of requirements to the missing alternatives. The primary basic alternatives do not always satisfy the participants of choice, but enable them to understand better what is missing, what is realizable in this particular situation and what is not. This class can be called the problems with constructed alternatives.

So, the alternatives appearing in the problems of decision making can be
(1) independent,
(2) dependent,
(3) predefined,
(4) appearing upon working out the rule of decision making,
(5) constructed during decision making.

We note that there are intermediate cases as well. Understandably, the nature and number of alternatives present certain requirements to the methods of decision making and offer one or another possibility. For example, if the alternatives are given, then it is advisable to apply to them the methods of data analysis (Bertier, Boutoche, 1975), make groups of dependent or similar alternatives, and establish the similarities and dissimilarities between the groups, and so forth.

If the number of alternative is small, then it is advisable to present them to the DM to analyze. Special attention must be paid to variants of presentations of alternatives such as holistic image, criterial representation, etc. If there exists a conceptual possibility of creating new alternatives, then concrete methods of their generation must be provided.

Criteria

Criteria provide a means for describing alternative variants of decisions and expressing the differences between them in terms of the DM preferences. The number of criteria in different theoretical constructions and decision making methods is usually more than one - probability and utility in the theory of subjective expected utility, effectiveness and cost in the "cost-effectiveness" method, etc.

The modern methods of decision making are oriented to allow for all distinctions of the alternatives and all riches of their descriptions, which approached formal schemes to the real life. Therefore, the multicriteria description of alternatives becomes during the last decade more and more popular. As a rule, at the beginning of problem analysis the criteria for estimation are not given and must be identified within DM-analyst interaction. If all or a part of alternatives are given, then criteria are identified by comparing them, as it is done in repertoire grids of Kelly (Kelly, 1955) and realized in the "MAUD" method (Humphreys, Wishuda, 1982). Here, criteria can be identified and scales of estimates constructed by analyzing the answers to questions "What is the difference between two alternatives?"

If no criteria are defined, then the DMs define them from their policies and requirements to the problem of choice. In doing so, consideration is given either to the preceding situations of decision making or to the anticipated alternatives.
Criteria can be mutually dependent or independent.

Criteria are referred to as dependent if the estimate of an alternative by one of them defines (deterministically or with great probability) the estimate by another criterion. For example, we can expect a luxury apartment to be expensive. The dependence between criteria gives rise to holistic images of alternatives which are meaningful for the DM.

The problems of decision making and methods for their solution depend on the number of criteria as well. For a small number of criteria varying between two and five, the problem of comparing two alternatives is fairly simple for the DM, qualities in terms of criteria can be compared in a straightforward manner, and a tradeoff worked out.

For many criteria, the problem becomes intractable. Fortunately, in such cases criteria can be merged into groups of dependent criteria. Such groups are usually independent having certain meaning, and their names can be regarded as the names of generalized criteria. Therefore, a hierarchy of criteria arises. In some problems, hierarchies of criteria can be constructed having different number of levels.

The possibility of identifying the pluses and minuses of an alternative, its advantages and disadvantages (for example, cost and effectiveness) provides a natural ground for grouping the criteria. Now, the advantages and disadvantages can be grouped again, for example, in criteria important for the DM and for active groups, long-term criteria, short-term criteria, etc. Dependence of criteria plays an important role in this grouping.

In a word, identification of the structure on a set of criteria makes the process of decision making much more intelligent and efficient.

Routine types of decision making problems

Depending on the requirements on the result of solution, the problems of decision making can differ substantially.

Ranking of alternatives. There are problems where it is required to establish a ranking on a set of alternatives. They are exemplified by a family ranking its future purchases in necessity, company heads ranking investment media, educaters ranking university graduates in their progress, and so forth. In the general case, the requirement to rank alternatives means that it is desired to establish relative value of each alternative.

Every so often one needs not a perfect ranking where all alternatives "parade" one after another and can be satisfied by a quasiorder where the alternatives are not all comparable. Here, some alternatives can be ranked in a fuzzy manner, that is, their position in a sequence can be defined by some interval - for example, an alternative can be worse than the tenth alternative and better than the fifteenth one, that is, have a rank within the interval between the eleventh and fourteenth alternatives. Isolation of the Pareto layers of alternatives is a well-known special case of qiasiorder. As will be seen in what follows, the construction of quasiorder requires much less information from the DM, but the quasiorder itself often can be a satisfactory solution of some practical problems.

Allocation of alternatives into quality-ranked groups. When buying an apartment or house, people usually group alternatives into deserving further detailed consideration, which requires efforts, and not deserving it. A physician examining patients can group them according to suspected diseases. A commodity expert can group commodities in terms of their quality. An entrant can appreciate educational institutions in terms of their prestigiousness. In the same manner people often group books, fashions, tourist routes, etc.

Generally speaking, grouping of objects is a characteristic human occupation, which is due to the fact that classification often provides a satisfactory solution to many practical problems, especially if the number of objects is rather great. For instance, there is no reason to rank strictly several hundreds of objects, but their grouping can give a quite satisfactory answer to the question about their quality. However, grouping is advisable also for a small number of objects if it is dictated by the informal formulation of the problem. We also note that grouping of objects is the outcome of work of many expert systems.

Choice of the best alternative. This problem was traditionally regarded as one of the challenges to decision making. It is often met in real practice. One is well acquainted with examples such as choice of one item in shop, choice of a job, choice of a design of a complex engineering facility. Furthermore, such problems are often met in the world of political decisions where the choice of alternatives is comparatively small, but the alternatives themselves are fairly complicated in terms of study and comparison. One may require, for example, the best variant of carrying out replacement of media of circulation, the best variant of land reform, etc. We note that many problems of political decision making feature construction of new alternatives in the course of their solution. The problem of choosing the best alternative usually arises if the number of compared alternatives is small and observable to the DM. In practical applications, there exists a significant difference between problems where alternatives are predefined and problems where the number of alternatives can vary on the course of solution (new alternatives can be added).

Economico-mathematical approach to problems of choice

The theory of decision making is regarded as an important section of economics because making of economic decisions is an important aspect of everyday human activity. As a consumer, the person must decide which commodities and for what prices are to be purchased. As a producer, the person must decide what finances must be invested into different projects, shares, or kinds of products. As a producer investing efforts and time into various kinds of activity, the person must decide how to reasonably lay them out. The economic theory found an answer to these matters by introducing the notion of utility (of commodity to the consumer) and the utility-price ratio.

It is assumed that people can measure (most often, subconsciously) values of various benefits, commodities, etc., in terms of the so-called utility.

Rational choice

Each portion of purchased commodity (e.g., bread or tea) has its consumer utility. The law of marginal utility reads as follows: the marginal utility decreases, that is, the subsequent portions of a commodity are less valuable to the consumer than the first ones, which is quite understandable from our everyday experience. If there exists a need for several commodities, the consumer attempts to allocate money so as to support constant ratio of the utility of a commodity to the general measurement unit (dollars, rubles, etc.). Stated differently, greater money must be invested into commodities of higher utilities. The same human behavior is characteristic of the problem of investments - more money are input into more useful areas of activity. The economists believe that this is the only correct behavior and refer to the person making such choice as *rational person.* .

It is assumed that rational persons are intrinsically consistent and that transitivity of choice is appropriate to them. For instance, if a commodity A is preferred to the commodity B and B is preferred to C, than A is always preferred to C.

Second, it is assumed that, when making a decision, the rational person maximizes the utility.

To conclude, what is done by the rational persons? First of all, they list all possible decisions and their consequences for which utilities (or money values, if possible, or determined. For each variant of decision, the probabilities of all its outcomes are determined (no matter how). Next, the expected utility of each variant is calculated by summing the products of utilities by corresponding probabilities. The best variant is that with the maximal expected utility.

The scientific foundation for the utility theory was laid by J. von Neumann and O. Morgenstern in their well-known "Theory of Games and Economic Behavior" (von Neumann and Morgenstern, 1947). The utility theory as presented in this book is axiomatic. The originators of the utility theory made use of the so-called lotteries, where two results (outcomes) with respective probabilities p and (1-p) exist, as simple problems of choice and demonstrated that if human preferences for simple problems (lotteries) satisfy some axioms, then the human behavior can be regarded as maximization of the expected utility.

The axioms used by the authors of (von Neumann and Morgenstern, 1947) assert, for example, that the person can compare all outcomes and is transitive, which it is possible to determine the probabilities under which lotteries constructed on pairs of outcomes (out of three) are equivalent, etc. The axioms are required to infer the theorem of existence of the utility function for a person that agrees with the axioms. The internal utility function is used to measure the utility of any outcome. The theory of the classic book of J. von Neumann and O. Morgenstern needs a quantitative measurement of all utilities and probabilities.

Subjective expected utility

The theory of von Neumann and Morgenstern assumes that the probabilities are given as objectively known magnitudes. D. Savage (Savage, 1954) developed an axiomatic theory enabling one to measure simultaneously the utility and subjective probability which gave rise to the model of subjective expected utility (SEU) where the probability is defined as the degree of confidence in fulfilment of one or another event. We note that this model is popular with many economists as a scientifically substantiated means for choosing the best decisions. The attractiveness of the SEU model can be accounted for its simplicity - only two parameters (utilities of the outcomes and probabilities of their fulfilment) must be measured. If events occur often, then the necessary measurements can be done.

There is another cause of popularity of the SEU model. As B. Fischhoff (Fischhoff, 1980) rightly notes the model allows one to select behindhand (after the choice) the parameters (probabilities and utilities) so that the SEU model can explain any choice. The theory of subjective expected utility is well protected from this point - if its predictions differ from the real human behavior, then one can always say that it would be well to use another function.

At one time the SEU model was regarded not only as a norm prescribing how to choose, but also as a description of human choice. Yet, there exists an extensive psychological literature on human errors made in assigning subjective probabilities (see below), and the SEU model became open to question.

Decision trees

The so-called decision trees (Raiffa, 1968; Brown, et al., 1974) became a very popular method of decision making in the 1960s and 1970s. Many complex decisions involve more than one choice. When deciding whether to develop an oil field, an oil company, for example, has to answer questions such as where and how carry out trial drilling, whether to invite consultants or confine itself to the available information, etc. The sequence of decisions can be conveniently represented as a tree.

The SEU-based method of decision trees perfectly described by Raiffa (Raiffa, 1968) is very popular. It decomposes the problem at hand into subproblems that are, in their turn, decomposed into subproblems an so on until the problem is represented as a decision tree. In some vertices of the decision tree decision is made by the DM, in other vertices decision depends on the objective or subjective probability of certain events. The decision tree terminates in outcomes to each of which a certain utility is assigned. The probability of an outcome is calculated as the product of subjective probabilities along the path going from the vertex of the decision tree. The outcome with greatest SEU is selected by "folding up" the decision tree from the end to the beginning, which defines the way of problem's solution.

Prospect theory

Attempts were made recently to update the utility theory so as to eliminate the most salient discrepancies between the theory and real human behavior. The prospect theory (Kaheman, Tversky, 1979; Currim, Sarin, 1989) is the most conspicuous attempt of this kind. By prospect is meant a game with probabilistic outcomes.

The prospect theory allows for three behavioristic effects:

(a) certainty effect, that is, the tendency to give greater weights to determinate outcomes,
(b) reflection effect, that is, the tendency to change preferences upon passing from gains to losses, and
(c) isolation effect, that is, the tendency to simplifying choice by eliminating the common components of decision variants.
All these effects being taken into consideration, the value of a lottery to gain the outcomes x and y with respective probabilities p and q is defined by

multiplying the utilities of outcomes by the subjective importance of the probabilities of these outcomes.

The weight function $\pi(p)$ expresses the subjective importance of the probability of a definite outcome. Although $\pi(0)=0$ and $\pi(1) = 1$, the function π has specific features near its extreme values. For instance, for small probabilities $p > \pi(p)$. We note that the ratio of values of π is closer to 1 for smaller probabilities than for larger ones.

The theory of prospects recommends to "edit" prospects before comparing them - for example, to eliminate identical outcomes with identical probabilities, to merge in one prospect identical outcome, and so forth.

Despite the fact that the theory of prospects eliminates some paradoxes of choice stemming from the utility theory (for example, the Alle paradox (Raiffa, 1968)), it does not eliminate all problems and paradoxes appearing upon studying the human behavior in the problems of choice. The possible paradoxes appear when editing the prospects. A quite natural desire to round the probabilities and utilities leads to different results of prospect comparison depending on the rounding (Wu, 1993).

The prospect theory, as well as the utility theory, relies on an axiomatic basis. A common problem with all axiomatic theories is validation of the axioms allowing one to use one or another form of the function of utility (value) of the theory.

Pareto set

The approaches based on the utility theory, subjective expected utility, and prospect theory regard utility (value) as a holistic estimate of an alternative. Their popular counterpoise is exemplified by the multicriteria description of alternatives where each alternative is characterized by estimates in many attributes, factors, or criteria. This description is often closer to the real world where each alternative has its pros and cons.

The first stage of analyzing multicriteria alternatives consists in isolating the Pareto set (in recent years, it is called more often the Edgeworth-Pareto set (Stadler, 1989)).

The Pareto set is known to include those variants of decisions that are not dominated by other variants from the point of view of the entire body of criteria. It is understood that the elimination of the variants not belonging to the Pareto set does not solve the problem of multicriteria choice, but only simplifies it. The problem itself of constructing and analyzing the Pareto set lies within the domain of applied mathematics where not a few interesting and useful results have been obtained. Yet, analysis of the Pareto set is only the first step toward solution. It is only natural to compare further the elements of the Pareto set, that is, to establish rules for choosing one of the objects estimated in many criteria.

Multicriterial utility theory

The next step in the evolution of the utility theory was marked by passing to the multicriteria utility theory (Keeney R., Raiffa H., 1976). The construction of a strict and harmonic mathematical theory of utility under multiple criteria was a great merit of R. Keeney and H. Raiffa. The theory also is constructed axiomatically, the general axioms of connectivity and transitivity on a set of alternatives, etc., being complemented by the axioms (conditions) of independence. There exist many such conditions (Humphreys, 1977) which conceptually define the possibility of comparing alternatives in some criteria, the estimates in other criteria being fixed (at different levels). For example, the condition of preference independence defines that comparisons of alternatives in two criteria are valid if the estimates in other criteria are fixed at any level. If this condition is met for all pairs of criteria, then the utility function is additive, that is, representable as a sum of one-criterial utility functions times the importance coefficients of the criteria. We note that since this definition refers to the case of certainty, the value function is used instead of the utility function. We also note that the multicriteria utility theory are vectored to the problems where many alternatives justify great efforts required to construct the utility (value) function.

The SMART method

Assuming that the criteria are independent and the utility functions are additive, one can substantially simplify the procedure of applying the method of the multicriteria utility theory. W. Edwards proposed a Simple Multiattribute Rating Technique (SMART) dwelling on these assumptions (von Winterfeldt, Edwards, 1986) which consists of

(1) computing the importance coefficients of criteria (see (von Winterfeldt, Edwards, 1986) for various methods of measurements),

(2) constructing the utility functions separately for each criterion, and

(3) calculating the utility of each alternative using the above approach.

Understandably, the SMART technique is fairly simple as compared with the approach of multicriteria utility theory, but the methods of multicriteria utility theory have more strict mathematical foundation.

Incomparability of alternatives. The ELECTRE methods

Another approach to comparing and estimating multicriteria alternatives, the ELECTRE methods appeared simultaneously with the theory of multicriteria utility. These methods, which were developed by a team of French researchers headed by Prof. B. Roy, are oriented to choosing from a group of alternative a subgroup of the best ones. They have the following two original characteristics.

(1) Criteria are regarded as persons (jurors) voting for one or another choice (B. Roy, 1985, 1996), which explains the special attention paid to each of the criteria whose weights as if reflect the degree of influence of each juror. If an estimate by one criterion is low, then the alternative has a serious defect (negative opinion of one of the jurors).

(2) The notion of incomparability of two alternatives is introduced. If the estimates of alternatives to a large measure are contradictory, that is, an alternative is superior in some criteria and inferior in the other criteria, then the contradictions do not compensate anyhow and cannot be compared. This notion is also of extreme practical importance because it identifies alternatives with "contrast" estimates which deserve special consideration.

The major ideas of the majority of methods from the ELECTRE family can be described as follows.

To each of the N criteria having numerical or qualitative scales, a number p characterizing its importance is assigned (in ELECTRE IV all criteria are equal in importance). B. Roy proposed to regard p as the "number of jurors' votes" for importance of this criterion. For any pair of alternatives a and b, a binary relation is constructed according to which a is superior to b under certain values of the agreement and disagreement indices that are defined as follows.

The agreement index (with a superior to b) is established from the fact of superiority of the weights of criteria in which a is superior or equal to b over the weights of criteria in which the estimate b is superior to the estimate a. The disagreement index is defined as a function of the most significant difference between the estimates b and a in the criteria where the alternative b is preferable.

In the ELECTRE III and IV methods, the agreement and disagreement indices are fuzzy, that is, fuzzy sets are used. The membership functions enable one to establish the degree of superiority (strong or weak) of one alternative over another.

In the ELECTRE methods, the binary superiority relation is defined in terms of levels of the agreement and disagreement indices. If the agreement index is above the given level and the disagreement one is below it, then the alternative a is declared to be superior to the alternative b. If for these levels alternatives cannot be compared, then they are declared to be incomparable.

It is important to emphasize that for given estimates of alternatives the given levels of agreement and disagreement, where alternatives are comparable, provide an analytical tool to the consultant who can investigate the set of alternatives by defining the levels and gradually reducing the required level of the agreement coefficient and increasing that of the disagreement coefficient. For each given pair of levels, a kernel of nondominated incomparable (or equivalent) elements is isolated. A smaller kernel can be extracted from it by varying the levels, and so forth. The analyst offers to the DM a whole range of possible solutions to a problem in form of different kernels. A single best alternative can be obtained eventually. The degrees of "violence" to the data characterize here the values of agreement and disagreement indices.

Method of analytical hierarchy

The method of analytical hierarchy (AH) proposed by T. Saaty has recently become popular in the United States and Britain (Saaty, 1980). It is based on a multicriteria description of the problem. Similar to the methods of the ELECTRE family, the AH method is oriented to working with a given (usually small) group of multicriteria alternatives from which best alternative must be isolated.

One can identify four main stages in the AH method.

(1) Structuring of the problem in form of level hierarchy - from objectives to criteria and from criteria to real alternatives. The elements of each level are listed exhaustively. Sometimes, intermediate levels (for example, of objectives) are introduced.

(2) The elements of each level are compared pairwise in the degree of their preference to the DM. A comparison language with nine degrees of superiority ranging from equivalence through weak superiority and so on to very strong superiority is introduced, and a numerical scale ranging from 1 (equivalence) to 9 (very strong superiority) is assigned to it, that is, to each verbal description a certain number is assigned. The results of pairwise comparisons are represented as a skew-symmetric matrix, and it's latent vector, which components characterize "on the average" the degree of superiority of each of the compared elements over other elements, are computed.

(3) At the lowermost hierarchical level, the real alternatives are compared pairwise in each criterion, that is, N comparison matrices, where N is the number of criteria, are constructed at this level.

(4) The index of value of each alternative is established using the method of weighted sums of estimates of criteria where the estimate (coefficient of superiority of the given criterion over other criteria) is multiplied by the weight (coefficient of superiority of the given alternative over the other alternatives in the i-th criterion).

We hold that the cause of popularity of the AH method lies not only in its simplicity, but also in that it enables the user to compare real alternatives separately in each criterion, which surely is of practical interest. At the same time, no method was recently discussed or criticized as widely as the AH method.

Of modifications improving some qualities of this method, it is pertinent to note the introduction by F. Lootsma of geometrical scale for translating qualitative comparisons into numbers (Lootsma, 1990) and the MACBETH method (Bana e Costa, Vansick, 1974).

Successive decisions

There exist a number of theoretical constructions concerned with the problems of making successive decisions, that is, decisions made at successive time instants. One of them is as follows: given are variants of decisions (estimated in one or more criteria) that are presented to the DM successively at random; needed is to choose the best one (the DM knows nothing about the quality of still not presented variants). This problem is known as "the problem of the best secretary". Methods enabling one to find the probabilistic estimates depending on some factors such as the number of variants, structure of preferences on the set of variants, etc., were developed (Dubov, et al., 1986).

Another formulation of the problem (von Winterfeldt, Edwards, 1986) is as follows: there are probabilistic quality estimates of decision variants and at subsequent time instants additional information arrives. The well-known Bayes formula is regarded as the optimal rule of variation of probabilities.

Repeated decisions

The "bootstrapping" method of R. M. Dawes (Dawes, 1988) deserves mentioning as a practical tool for making repeated decisions that works well in practice.

In situations where the objective consequences of many repeated decisions are known, one can select a model predicting to the best advantage the outcomes of a decision. This can be done in various ways. Not claiming by any means to describe the real DM behavior, R. M. Dawes proposed to employ a simple linear model of weighted sums of criterial estimates and demonstrated experimentally that the model has good accuracy of predictions.

Of interest are the results of the following experiments. Using the same linear model, weights were reconstructed from a half of sample. Predictions made from these weights were compared with opinions of the DM using the same criteria. Weights were either assigned by the DM or taken to be equal. Better accuracy was obtained namely in the latter case, that is, for rough linear models where weights were established in an overtly nonoptimal manner. (One of the papers of R. M. Dawes is named "The robust beauty of improper linear models in decision making" (Dawes, 1982)).

Mathematical modeling in problems of choice

The economico-mathematical approach to choice focuses on constructing DM's utility function or decision rule. Here, as if a model of DM's subjective estimate of the problem of choice is constructed.

In the domain of decision making there exists traditionally another line of research using mathematical models of physical systems often including human teams. It stems from the operations research.

Operations research

The operations research as practical discipline dates back to the World War II. It was intended to construct models of military operations and find with their help the best decision.

By operations research currently is meant the application of quantitative mathematical methods with the aim of substantiating decisions in all areas of purposeful human activity (Ventsel, 1972). Construction of a quantitative model, choice of the optimality criterion, and determination of the optimal decision are the main means for solving any problem of operations research. The literature on operations research presents excellent examples of its practical application.

The following feature of operations research as an already set line of research must be noted.

1. The models used are objective. Within the framework of operations research, the construction of models is regarded as a means of reflecting the objective reality. Many experts in operations research emphasize that the construction of models and choice of parameters are not simple matters. It is assumed that everything depends on expert's experience and skill. Ignoring the computational burden, the optimal solution can be found in a unique manner if a model correctly reflecting the real world is constructed and the criterion established. Interestingly, G. Wagner noted (Wagner, 1969) that different experts relying on the same data must come to the same results. The justness of this requirement is quite understandable if we turn to the classic problems of operations research such as inventory control, strategic games, queuing theory, etc.

We describe such a problem by way of example. The workers of a factory manufacturing different types of products from time to time must change tools required for work. The tools are received from storemen. If the number of storemen is low, then the workers wait in queues and lose time. If there are many storemen, then they lose their time. It is required to determine the number of storemen distributing tools so that the time losses of workers, on the one hand, and storemen, on the other hand, result in the minimal production losses. The criterion is defined in the formulation itself. To solve problems of this kind, the distributions of probabilities of calls for tolls are studied, queuing times are evaluated, a model of production cost is constructed, and with its help the optimal solution is found.

Obviously, different skilled researchers come to the same result. From this point of view, the approach of operations research to modeling is no different from physics or other natural sciences. Indeed, the flow of workers to the storemen does not differ from the flow of products to checkpoints with limited throughput or flow of particles to detectors with limited speed of registration. The objectiveness of model in these examples is apparent.

2. The problems of operations research are solved at request of management

An analyst getting such an order studies the system and the environment and tries to establish an adequate model. At this stage, the managers themselves are not usually required. The abundant descriptions of applications of the operations research methods emphasize that teams of analysts themselves find successful solutions. Of course, sometimes managers can provide additional information, but their role does not differ from the role of any executive of the organization. As soon as a solution is found, the task of managers is to realize it. Stated differently, managers give orders and get ready solutions, the rest being done by the experts in operations research. In the general case, the order can be formulated as follows: needed is to find the best (optimal), unique, and scientifically substantiated solution. When the managers define such a order, they are in an advantageous position - they rely on the efficacy of scientific approach.

3. There exists an objective criterion of successful application of the methods of operations research. If a problem under consideration is obvious and the criterion is defined, then the analytic method immediately shows to what extent the new solution is superior to the older one. If in the above example the

optimal number of storemen is found, then there is no reason to hire more or less persons.

"Cost-benefit" and "cost-efficiency" analysis

The classic methods of operations research - primarily, mathematical programming - were oriented to finding the optimal solution through a sufficiently reliable mathematical model. Further attempts of expanding the domain of operations research has led to a group of practical problems of allocating money to the public sector where models often are subjective and criteria multiple.

To compare variants of solutions in such problems, the "cost-benefit" method, which is oriented to monetary evaluation of the social benefits, was proposed. It is fairly simple. First of all, importance coefficients are assigned to all objectives (problems). One objective has a monetary evaluation. The monetary evaluations of other objectives were determined from it by proportional modifications in accordance with the relations of importance coefficients (Hinrichs, Taylor, 1969).

For case where efficiency cannot be expressed in monetary terms, the "cost-effectiveness" method was developed. For the first time, this method was used to analyze military engineering designs (Houston, Ogawa, 1966).

The "cost-effectiveness" method consists of the following three main stages:
- construction of the model of effectiveness,
- construction of the model of cost,
- synthesis of evaluations of cost and effectiveness.

We note that in terms of objectivity models of cost and effectiveness were close to the models of operations research, but their output parameters cannot be united by means of a given dependence. Therefore, the opinion of the manager defining the extreme values of cost and desired values of effectiveness is often used. The ratio of cost to effectiveness is often used, but it is recommended to pay attention to their absolute values. The main difference from the approach of operations research lies in the appearance of subjective judgments upon synthesizing cost and effectiveness.

It is generally recommended to employ the following two approaches at the stage of synthesis of evaluations of cost and effectiveness:

(1) fixed effectiveness under minimal possible cost where the "cheapest" of the alternatives having the required effectiveness is chosen or

(2) fixed cost and maximal possible effectiveness (the case of budget constraints) where the "most effective" alternative satisfying the cost constraints is chosen.

Quite evidently, the aim of these approaches is to translate into constraint one of the criteria for evaluation of alternatives. In doing so, we have a typical problem of operations research; nevertheless, there, undoubtedly, remains a degree of subjectivity in determining the constraint on one of the criteria.

Two stages of problem solution

Multiple criteria and subjective DM preferences by no means corresponded to the approach of operations research and acquirements of mathematicians using it. That is why it was suggested to isolate two stages in the solution of a decision problem.

The first stage is objective analysis of the problem and identification and investigation of the apparent relations. The second stage, final determination of the best solution with regard for many factors, is carried out by the manager who is provided with the results of the first stage. The usual results of the first stage are represented by an objective model and the Pareto set. Then, a list of variants and their estimates is presented to the DM who examines them and chooses one variant.

As it turned out later, the task of choosing among multicriteria alternatives proved to be a psychological challenge to the DM. Attempts to simplify it by introducing importance coefficients for the criteria and by "folding" the criteria into one global criterion also required that persons execute complicated cognitive operations.

Only gradually an appreciation came of the fact for the DMs the multicriteria problems with objective models represent a special class. Upon solving them, they simultaneously examine the domain of admissible decisions and work out a tradeoff between the criteria.

What major challenges are presented to the DM by mathematical modeling? When using a mathematical model, the DMs must work out an understanding of what levels of quality and by which criteria are simultaneously attainable. To this end, they must investigate the model's response of to one or another of their actions. As a rule, the model is described by many variables and is difficult to grasp; but nevertheless, the DMs determine the "limits of possible" and simultaneously work out acceptable tradeoffs between the values of various criteria by studying responses of the model to their actions directed at attaining one or another values of criterion. An entire class of man-computer procedures aiding DM to solve such problems came into existence.(Steuer, 1986).

Multicriteria counterparts of the well-known problems of operations research

The introduction of multiple quality criteria enables one to obtain multicriteria counterparts of the well-known problems of operations research. First of all, we mention the multicriteria transportation problem. Additional criteria are readily built in into the generalized transportation problem (Wagner, 1969) which can be formalized as a multicriteria problem of linear programming for which multitude of methods was developed (Steuer, 1986).

The assignment problem exemplifies the classic problems of operations research (Wagner, 1969). If the persons to be assigned and possible jobs are evaluated in terms of many criteria, the multicriteria assignment problem arises. The multicriteria assignment problem was discussed in some publications (Larichev, 1987). It can often be formalized as the multicriteria problem of integer programming.

The bin packing problem is one of the popular problems of operations research (Garey, Johnson, 1979). If the packed objects are estimated in multiple criteria, then a multicriteria packing problem appears whose algorithm of solution is described in (Larichev, Furems, 1985).

Man-computer procedures for solving multicriteria problems of mathematical programming

There exists a great deal of man-computer procedures enabling the DM to examine the domain of admissible decisions and at the same time to establish a compromise between the criteria (Steuer, 1986; Larichev, 1987,Lotov,1997).

The man-computer procedure consists of alternating phases of analysis (performed by the DM) and optimization (performed by the computer). A phase can consist of more than one step.

Optimization phase (computer):
- using the information received from the DM at the preceding step, a new decision is computed and auxiliary information for the DM is generated.

Analysis phase (DM):
- the presented decision (or decisions) is estimated and its admissibility is determined. If the answer is positive, then the procedure terminates; otherwise, auxiliary information is considered;
- additional information is communicated to enable computation of a new decision.

The man-computer procedures differ in content and execution of the above steps. Their efficiency depends mostly on the nature of DM-computer interaction which is represented in terms of quality and quantity of information. The recent studies (Larichev et al., 1987) are oriented mostly to the behavioral aspects of DM-compufer interaction.

Various kinds of problems

Attempts to apply the operations research to administrative systems have demonstrated that the nature of problems under consideration differs significantly. These differences were first noted by H. Simon and A. Newell who suggested a lucky classification of the problems (Simon, Newell, 1957) into three classes.

1. Well-structured or quantitatively formulated problems where the essential dependencies are clarified so well that can be expressed in numbers or symbols that eventually are evaluated numerically.

2. Ill-structured or mixed problems having both qualitative and quantitative elements; here, the ill-known and indefinite qualitative aspects of problems have a tendency to domination.

3. Unstructured or qualitatively formulated problems describing only the major resources, attributes, and characteristics whose quantitative interrelations are absolutely unknown.

According to this classification, the typical problems of operations research can be regarded as well-structured.

The "cost-effectiveness" and "cost-benefit" methods were among the first attempts to compare the variants of solution of ill-structured problems. Similar to the methods of linear programming as used in the operations research, the "cost-effectiveness" and "cost-benefit" methods became regarded as the set of system-analysis tools for ill-structured problems.

System approach to problems of choice

The economico-mathematical approach and that of mathematical modeling are aimed at developing particular methods for comparing alternatives. In real situations, however, the alternatives occur while analyzing the problem and not from vacuum. The point is how to carry out this analysis, how to pass from the DM's drives and general objectives to the solution of the problem?

The system approach strives for answering this questions. As is shown in what follows, the variants of the system approach focus on the stages of investigating the problem, rather than on the methods of comparing the alternatives (Churchman, 1969).

System approach

The word "system" is used today as an adjective in diverse combinations. In engineering systems they speak about system engineering; there exist terms such as system analysis, system management of projects, system design of organizations, etc. Obviously, in this context the word "system" relates to the ideas of the general system theory and cybernetics.

The main notions of the system approach such as system, process, input, output, feedback, and constraints are applied to system of most different nature. In cases we are concerned with, we may analyze the processes of choosing a design of a unique object, developing rules for bank decision making, etc. By observing these processes, one can identify appropriate systems and their subsystems, gain an insight into their relationships with other systems, determine the input (input information), output (decision), feedbacks (analysis of the decision), and constraints (resources, work force, etc.).

What is usually meant by the words "system approach"?   To answer this question, let us discuss the available recommendations for solving problems of different nature.

The following five stages are usually isolated in the approach of the system analysis and operations research:

(1) identify the objective(s),
(2) identify alternatives means for attaining the objective,
(3) determine the resources required by each system,
(4) construct a mathematical model, that is, a number of dependencies between the objectives, alternative means to attain them, environment, and resources,
(5) establish the criterion for choosing the preferable alternative.

In many papers and books dealing with successive approach to complicated systems one can meet similar stages.

We note that different system approaches have some points in common: definition of a clear-cut sequence of actions, consideration of objectives and means, identification and successive consideration of alternative variants of problem solution, and desire to make a rational choice between them.

In considering a problem, H. Simon identifies three stages (Simon, 1960) such as search, design, and comparison of variants. As can be seen, the system approach is oriented to the first two stages. It was conceived as a "metameans" for problem solution embedding concrete methods for comparing alternatives.

Thus, the system approach to most diverse problems is based on isolating the system from the environment and determining a set of successive consistent stages of considering the problem. We refer to these features as general scheme of the system approach (Larichev, 1979).

What is the difference between the system approaches to different problems? It lies, primarily, in the means of analytical comparison of the alternatives.

Systems analysis

One of the fathers of systems analysis, A. Enthoven defined it as a reasonable approach to decision making which is accurately described as the quantitative common sense (Enthoven, 1969).

The word "quantitative" underlines the fact that, in formal terms, the first variant of systems analysis is a combination of the general scheme of systems approach and the above "cost-effectiveness" method for comparing alternatives. It is precisely this method that requires quantitative representation of variables. The systems analysis was used mostly for analyzing and solving problems of large organizations such as departments (Novik, 1965) and large companies. The desire to structure correctly the problem objectively led to centralization of the decision making. Undoubtedly, there was a great need for such a centralization and the systems analysis enabled one to satisfy it.

The studies of economists and psychologists provided an insight into human decision making in large organizations. Ch. Lindblom notes the officers of organizations try to make as small changes in the existing policy as possible to be able to adjust to the environmental changes (Lindblom, 1959). It is not only easier to work out such changes, but also to coordinate them within an organization. The sequence of changes is mostly the means for forming the current policy. Lindblom also believes that this way of solving problems is more realistic because it requires less efforts and is more customary for the managers. On the other hand, this approach is more conservative and is not adjusted to dramatic changes in policy.

Similar discoveries were made by H.Simon (Simon, 1960) who introduced the notion of satisfactory decisions as a counter to the optimal ones. In organizations, the life itself brings people to seek satisfactory decisions - the environment is too complicated to be described by a model, the multiple criteria are defined incompletely, there are many active groups influencing the choice, etc. This natural behavior of the personnel resulted in that the strategic objectives were lost amid the petty, everyday routine.

The systems analysis was an attempt to relate the current decisions and strategic objectives. Within the general framework of systems approach, the so-called trees of objectives were constructed, and a gradual passage from objectives to means was made. This approach was especially popular in military decisions (Quade, 1963).

Criticism of the systems analysis

After the craze for the systems analysis, it was subjected to a biting criticism (Hoos, 1972) for some substantial procedural defects that were made manifest upon passing from the military planning to planning of in civil agencies. Interestingly, the most severe criticism was directed against the inadequate and (obligatorily) quantitative representation of the objectives, as well as against the "cost-effectiveness" method.

Stated differently, the union of the general scheme of systems approach and the "cost-effectiveness" method was regarded and criticized as an organic whole. The criticism was aimed at the quantitative indicators reflecting only a part of the contents of strategic problems under study. It was no accident that the indicators were quantitative - only in this form they can be used in mathematical models and the "cost-effectiveness" method.

For example, Ida Hoos writes about the deliberate deception of public opinion by the Planning-Programming-Budgeting (PPB) system creating an illusion that military conflicts can be solved only by military and not political means (Hoos, 1972). She quotes the then-Secretary of Defence Robert MacNamara who declared in 1962 that all available quantitative measures were indicative of the American victory in Vietnam. It is understood that the quantitative measures are a poor reflection of human desire to fight and sacrifice themselves.

Along with the criticism of the "cost-effectiveness" method as being inadequate to qualitatively stated problems, the general scheme of the systems approach also was assaulted. In their convincing paper, H. Rittel and M. Webber described the problems of planning as wicked ones for which the oriented succession of the stages of the systems approach loses any sense (Rittel,

Webber, 1973) - objectives depend on means, and means are influenced by the formulation of objectives. Solution of social, urban in particular, problems depends on nontrivial ideas which sometimes change completely both objectives and means.

"Soft" systems analysis

A new version of the systems analysis was suggested by P. Checkland (Checkland, 1981) who introduced, in contrast to the well-known variant of systems analysis (Quade, 1963), the notion of "soft" systems analysis. The main vector of his research is the development of the systems approach as such, use of feedback for studying problems, and elimination of the "cost-effectiveness" analysis as a "rigid" means not corresponding the nature of the problem. The "soft" systems analysis offers the researcher a set of logic stages in the study of problem. Determination of the problem "roots", that is, its main causes, is one of the major stages. The researcher applying the logic stages moves toward the solution. Yet, failures are possible along this way. Then one must return to the beginning and look for new problem "roots".

Thus, Checkland proposes a closed sequence of the stages of investigating a problem (search of causes - objectives - means - search of causes) which is mush more realistic. Indeed, structuring of this problem is a creative task. The main concern of the researcher at the initial stages of solution is to formulate the problem in a new, "solvable" form.

It is well to note that the "soft" systems analysis is oriented to improving the art of analysts considering real problems. We think that it is a means for gaining insight into a problem, rather than a solution, though there exist examples of problems solved to advantage.

We note that "explaining models" are constructed for better understanding by many active groups (Ostanello, 1990). They provide a means for reaching a coordinated opinion on a problem and, at the same time, indicate to a "solvable" formulation of objectives.

Computer approach to the problems of choice

The proliferation of computers in various domains of human activity reached the domain of human decision making.

Initially, the computer was regarded as a means for "mechanizing" the methods of decision making and fast execution of computations, but later the situation changed and it became an independent person, partner of the man in decision making. And here two lines of study can be identified.

Decision support systems

The notion of decision support system (DSS) was given in (Bonczek et al., 1981; Olson and Courtney, 1992). We refer the reader to the books and survey (Larichev, Petrovsky, 1987) for a detailed overview of DSSs, and here we present only the main definition and some information. By decision support systems are meant man-computer systems enabling the decision makers to make use of data, knowledge, and models, both objective and subjective, in analyzing and solving ill-structured and unstructured problems.

The conceptual DSS model consisting of the following units:
(1) man-computer interface,
(2) problem analyzer,
(3) decision making unit,
(4) database,
(5) model base,
(6) knowledge base.

DSSs are usually constructed for a certain class of problems to help the DMs in analyzing them. The DMs request necessary data, get from DSS advice on the experience gained in solving similar problems, and try to make use of various methods and expert knowledge to get a solution. Such a profound analysis depends primarily on inputting initially in the DSS the necessary data, knowledge, and methods. It allows the DMs to gain an insight into the problem, to specify their preferences, and to work out the best variant of its solution.

The majority of existing DSSs does not have the exhaustive set of the above units and is oriented to a relatively narrow range of problems. Today, the DSSs develop along the following paths:

- merging DSSs with management information systems and communication systems,

-combining DSSs with expert systems and creation of "intelligent" DSSs, and

- improvement of the DSS technology (personal computers, sophisticated software, friendly interfaces).

Before the coming of such DSSs, we expect that systems that would be able to adjust to the way of thinking of a person, to simulate his/her working methods, and to become his/her alter ego will appear. Yet, some basic boundaries exist here - the DSS per se cannot generate a qualitatively new variant of decision, although there is hope that it can be established either through man-DSS interaction or guessed owing to this dialogue. We think that this expectation alone justifies the efforts aimed at developing increasingly efficient decision support systems.

Expert systems

Expert systems exemplify another domain of application of computers to decision making. They are design to store in computers the knowledge of skilled experts so that it could be used by specialists of much lower qualification. Stated differently, the expert systems are oriented to the class of problems with repeated solutions where expert's experience and intuition grow with years. It is assumed also that the problems to be solved are ill-structured. The would-be users of expert systems must want to get advice of experienced experts. Examples of such systems are manifold - from medical and technical diagnostics to geology and so forth. The human intuition is especially valuable in solving ill-structured problems. Guesses relying on the previous experience and flair of expert enable him/her to solve problems at an amazingly high level. This fact suggested the idea of transferring these abilities to the computer.

Many recent books and overviews (see, for example, (Hayes-Roth, et. al., 1983; Codier, 1984)) discuss expert systems and methods of their design. Below we list the main standard units of the majority of expert systems:

Expert interaction unit: the expert knowledge is input into the computer, handled, and added to the knowledge base through this unit.

Knowledge base: stores the expert knowledge input into the computer.

Database: stores the data about the knowledge domain, problem structure, cause-effect relationships, etc.

Logic inference unit: enables the user to make use of the expert knowledge by inputting into the expert system the description of a particular situation in response to which the mechanism of logic inference finds appropriate expert knowledge.

Explanation unit: provides the user with explanations of the logic of expert's reasoning.

The leading experts in artificial intelligence agree that the knowledge bases are the bottleneck of expert system design. An approach to this problem was proposed in (Larichev et al., 1991).

Psychological approach to the problems of choice

Although human thinking is the main subject matter of psychology, studies of human decision making appeared comparatively recently. The major line of psychological studies was represented until the 1950s by the behaviorism which reduced human behavior to the well-known "stimulus-response" chain. Behaviorism was replaced by the cognitive psychology which today remains the main domain of research.

The studies of human information processing system represent one of main problems of the cognitive psychology. Numerous experiments carried out by psychologists in various countries over the recent three decades provided abundant data about human brain, perception, and memory. We focus on the studies related to the problem of human decision making.

Model of memory

The three-component model of R. Atkinson and R Shiffrin seems most interesting and plausible (Atkinson et al., 1974; Atkinson et al., 1993). In our opinion, its advantage is the fact that it adequately explains the experimental data on human data processing (see the overview (Simon, 1991)). According to this model, there exist three kinds of memory - sensor memory, short-term memory, and long-term memory - differing in time and amount of stored information, method of coding, and level of organization of the stored information. External information goes to the sensor registers where it is kept for one third of second and then is transmitted to the short-term memory where it is coded and can be stored for up to thirty seconds (or much more in case of repetitions). Information further is either erased or sent to the long-term memory which can be imaginated as a store of unlimited volume where information can stay for infinite time.

This and other models stem from the so-called computer metaphor which draws the analogy between the computer (data input, short-term memory, and external memory devices) and the structure of operation of the human brain. Despite its simplicity, the computer metaphor proved to be amazingly fruitful in explaining the results of various psychological experiments.

Short-term memory

The majority of psychologists agree that the processes of human decision making take place precisely in the short-term memory. According to the model, it is updated by information both from the environment (through the sensor memory) and the long-term memory. The content of the short-term memory sometimes is identified with the contents of conscience because the man controls operations over the information in short-term memory. The most important characteristic of short-term memory is its size that is defined by the number of simultaneously stored information units. The main conclusion of the authors of various studies is that the size of short-term memory is limited.

In his famous paper about the magic number 7±2. J. Miller generalized numerous experimental studies of the human ability to process information and discern levels of stimuli (Miller, 1956; Baddley, 1994) and concluded that the short-term memory is limited in the number of stored units. Miller called this unit as chunk. The number of chunks in most diverse experiments did not exceed the magic number "7±2", chunk being either letter of phrase - anything perceived by the subject as one notional image. For example, the typist memorizes at most seven letters; but in other memory tests chunk can be a complex logic image.

H. Simon studied in detail the question of chunk on himself (Simon, 1974) by memorizing words and phrases having different notional contents and related in different ways. The results mostly corroborated the findings of Miller. The time of learning was shown to be dependent also on the number of chunks.H.Simon concludes that the psychological reality of the chunk is rather well demonstrated and that the size of short-term memory varies from five to seven chunks.

Thus, in the short-term memory information is represented in form of chunks. How does the man use this information in decision making? This question is partially answered by the experiments of Sternberg (Atkinson et al., 1993) where the subjects memorized sequences of figures not exceeding the size of short-term memory and then were given a figure and asked if it is present in the sequence. Judging from the experimental results, the subjects behaved as follows: they compared successively the given figure with all figures in the sequence and then decided whether it was in the sequence. This strategy is advisable if the time of decision making is much longer than that of comparing. Sternberg determined the time of one comparison as 35 milliseconds.

Our experiments concerned the effect of limited short-term memory on the results of classification carried out by the subjects (for more detail see Ch. 5).

In these experiments the subjects classified multidimensional objects by assigning them to one of several classes of decisions. Each object was characterized by a set of attributes measured by ordinal scales with verbal estimates. The experiments have demonstrated that the behavior of the subjects (number of errors and contradictions and complexity of the boundaries between classes of decisions) depend on the following three parameters of the problem: number of attributes, number of estimates on the scales, and number of decision classes. A domain was found in the space of three parameters where the subjects were able to cope well with the problem. If the problem parameters went outside the admissible domain (for example, there were more attributes or decision classes), then many contradictions were observed.

A detailed analysis of demonstrated that for classification the subjects used structural units of information (combinations of estimates of attributes) to define the boundaries between classes.

Presumably, they placed a certain number of chunks in their short-term memory and compared with them the description of some object in the language of attributes. As soon as the size of a problem exceeded the boundaries of the admissible domain in the parameter space, some subjects tried to use more than nine units of structural information (as was revealed by the analysis of their answers). In doing so, the number of errors grew rapidly because the current structural object did not correspond to the structural units of information in the short-term memory. For diverse (both in meaning and combinations of parameters) tasks that were coped well by the subjects, the number of structural units of information (chunks) did not exceed nine.

Long-term memory

Although decisions are made mostly in the short-term memory, both memories exchange information continuously. And generally speaking, their relations are very strong. There exists an opinion that in fact they are not different neural systems, but rather correspond to different activation states of a unique neural system.

There are many different and fairly complicated models of the long-term memory corresponding each to certain portion of experimental data. It is our belief that for decision making the most promising model is that based on semantic closeness (Klatzky, 1975) where a semantic class can be represented as a set of attributes in the long-term memory. Each object is representable as a point in the space of attributes, short distances corresponding to near objects. As Simon noted, the long-term memory resembles a large encyclopedia, and the experts have an extraordinary ability to perform fast indexed accesses to their knowledge encyclopedias.

Human heuristics and biases

As shown above, the economico-mathematical methods are largely based on human evaluations of the probabilities of various events. Therefore, the entire research of the theory of decision making methods was greatly influenced by the psychological studies of P. Slovic, A. Tversky, B. Fischhoff, et al., who proved that the human errors made upon evaluating the event probabilities are biased. The main causes of these errors can be represented as follows (Kaheman, Slovic, Tversky, 1982).

Judgment from representativeness: people judge about the membership of an object A to the class B only from its similarity to the typical representative of B disregarding the a priori probabilities.

Judgement from availability: people often evaluate the probabilities of events on the basis of their own meeting with such events.

Judgement from the anchoring point: if the initial information is used as the reference point for determining the probabilities, then it exerts significant influence on the result.

Superconfidence: people place too much confidence in their evaluations of event probabilities.

Tendency to eliminate risk: people try as much as possible to eliminate risky situations.

The results of these works question the possibility of practical application of many of the above methods. This problem was the subject matter of many lively discussions and give rise to "two camps in rationality" (Jungermann, 1983). These psychological studies played undeniably a very important role in that they drew attention to the limitedness of the human information processing system.

Descriptive studies of multicriteria problems

The multicriteria problems of decision making represent an especially complicated class of problems facing the human information processing system. Multiple criteria overburden the human memory and compel the man to have recourse to various heuristics in order to cope with problem under a limited size of the short-term memory.

Yet, the multicriteria problems occur more and more often in human activity, which is due to the need to allow for many different factors. That is why in the last decade the psychological studies of human behavior in multicriteria choice are carried out actively. Special methodologies of these studies have been developed (Harte et al., 1994; Russo and Rosen, 1975).

What are the results of studies?
Most often, consideration was given to the following problems:
(1) choice of one or more best alternatives from the set and
(2) estimation of an individual alternative (evaluation of utility, attribution to one of several classes of decisions).

Numerous studies indicate that there exist two main groups of human strategies used in these problems.

1. Compensation strategies where the subjects try to compare the evaluations of one alternative with those of another alternative.Main strategies of this kind were found: (i) determination of the utility of each alternative with subsequent comparison and (ii) comparison of evaluations of alternatives separately in each criterion and summation of the differences. They are known, respectively, as the additive model and the model of additive differences.

2. Strategy of elimination where the subjects eliminate the alternatives not satisfying one or more criteria. There exist two strategies of this kind: (i) elimination by a combination of evaluations in some criteria and (ii) successive elimination by individual criteria, which corresponds to the model of elimination by aspects.

The results of numerous studies provide an insight into the effect of such a specific feature of the information processing system as the limited size of short-term memory on human behavior in multicriteria problems. Upon using one of the compensation strategies, namely, that of determining the utilities of individual alternatives with their subsequent comparison, the man must take into consideration all evaluations of alternatives, which is extremely difficult. Therefore, as the experiments showed, people have recourse to a simplifying technique of just calculating the number of criteria by which one alternative is superior to another and take the alternative with the greatest number as the best one (heuristics A).

Upon using the second compensation strategy, that of summation of the differences of evaluations of alternatives by criteria, people often disregard the criteria for which such differences are small (heuristics B).

Although a precise elimination strategy would be to determine the combinations of criteria evaluations that correspond to certain classes of decisions, people often make use of a simpler strategy of elimination by successively considered criteria (heuristics C).

As studies have shown, the heuristics A, B, and C and their combinations are the main heuristic rules used in human comparison and evaluation of multicriteria alternatives.

Theory of search of the dominant structure

Relying on a number of studies, a general cognitive theory of choice, the theory of search for dominant structure was proposed (Montgomery, Svenson, 1989). In terms of G. Montgomery and O. Svenson, the DM tries to create the so-called dominant structure - such a representation of the problem that makes it clear that (a) an alternative is superior to all other alternatives by at least one criterion and (b) all disadvantages of this alternative as compared with other alternatives are compensated or eliminated in one or another way.

The process of decision making consists in searching for and constructing the dominant structure. When this goal is attained, the DMs know that they can explain their choice.

The experiments suggested that at the initial stages of decision making the DM hypothesizes that one of the alternatives is a candidate for final choice. This alternative must be suitable for building around it a dominant structure. If the DMs find discrepancies between the constructed dominant structure and the presumably best alternative, they return to the stage of constructing the dominant structure in order to eliminate the discrepancies.

Theory of constructive processes

J. Payne suggested and substantiated another theory of human behavior upon choosing the best multicriteria alternative(s) which can be called the theory of constructive processes.

When comparing multicriteria alternatives, people can use various strategies (see above). The studies of J. Payne (Payne, 1993) have demonstrated that in the process of decision making the subjects often choose a strategy depending on the specific features of alternatives under consideration (their evaluations by criteria). Here, the human preferences of alternatives and criteria are very unstable. At the local stages of comparison, rules (or their parts) can vary depending on the relation between the required human effort and accuracy of choice. Strategies often can be chosen erroneously in response to one or another (often insignificant) characteristic of alternatives.

As J. Payne notes, this behavior is characteristic of untrained subjects. People experienced in decision making , as well as regular decision makers have their preferable strategies for solving problems.

Mutual influence

The cross-influences of the above lines of research become more and more manifest in the course of the recent years. This is not surprising because the same subject of study compels one to seek both better means for solving the problem and more plausible explanations of the observed phenomena.

First of all, we should mention the influence of psychological studies on the economico-mathematical and model approaches. The desire to evaluate the developed procedures by psychological and sociological criteria is observed in increasing frequency.

More and more attention is paid to the primary analysis of problem and its structuring. Publications appear asserting that much depends on correct formulation of the problem and that the analyst's art plays here the major part. The "soft" system analysis also focuses on problem structuring and correct formulation.

The progress of microprocessors enables one to create increasingly perfect decision support systems realizing the idea of man-computer symbiosis.

The question of deep causes of human behavior in decision making comes to the foreground. Numerous psychological experiments enabled one to elucidate many characteristics of the human information processing system, but much remains indefinite.

Many recent discussions were devoted to the advantages and disadvantages of different methods of decision making. Their usefulness is obvious. Yet, one must keep in mind that each method of decision making has a preferable domain of application. It is unlikely that the authors of methods and theories would insist on universality of their approaches.

Construction of methods for new, previously unexplored classes of problems (such as, to the author's opinion, unstructured problems having only qualitative parameters) remains important.

We believe that the results of studies discussed in this overview prove that the conditions have already been created for a multidisciplinary synthesis of knowledge about the human decision maker. This synthesis gives rise to a new approach to constructing methods of decision making.

2 A NEW APPROACH TO UNSTRUCTURED PROBLEMS OF DECISION MAKING

Unstructured problems with qualitative variables

In the H. Simon's classification of problems presented in chapter 1, one can mark off a class of unstructured problems featuring variables of qualitative nature and unknown relationships between the variables. H. Simon noted that, although one often faces such problems when making strategic decisions of economic, political, etc., nature, the means for solving them were developed inadequately. In fact, the main characteristics of these problem are of qualitative nature, and sufficiently reliable quantitative models are still lacking. Let us consider some detailed examples.

1. A statesman chooses ways for providing his country with food. The main national producers of food are inefficient collective enterprises holding land and monopolizing (until recently) the market which is now captured by foreign producers of food. More efficient private farms are scanty. On the one hand, the statesman is pressed by a strong agricultural lobby

representing the collective enterprises demanding grants-in-aid and high import taxes. The partisans of collective enterprises advocate "preservation of national food production" as the goal of state policy. On the other hand, the statesman is under pressure of consumers who, naturally, are interested in lower prices and abundance of products and support free market competition. They regard "providing the population with food for reasonable prices" as the goal of state policy. As we can see, the active groups force different goals and ways of action on the DM.

Any process of working out a policy requires that the DMs carry out discussions with their followers and opponents and try to convince them, which is usually done by arguing in a language understandable to all sides. The DMs who know what they need are extremely sensitive to slightest changes in the formulations of their goals and criteria. To convince their entourage that their actions are consistent and well thought-out, they also must verbalize any tradeoffs between criteria.

2. The president of a large company decides to construct a large enterprise such as nuclear power plant, hydroplant, gas pipeline, etc., and has to evaluate the demand for this product with regard for the actions of possible competitors and the payback time, as well as the risk of investing into a particular region (the more so, a country). Consideration should be given to the arguments of the environment defenders who must be convinced that the would-be enterprise is not adverse to the interests of population and nature. The probabilities of accidents and their consequences for population and environment must be evaluated. It is desirable to demonstrate to the local authorities the benefits such as workplaces, payments to the local budget, etc., that will be brought about by the new enterprise.

Understandably, the discussion between the president, his/her assistants, and active groups goes on in a natural language. By the time of decision making, many factors are indefinite, but one must be able to estimate the degree of indefiniteness and convince other people in consistency and correctness of this estimate.

3. A large foundation supporting scientific research and development is set up, and one has to work out the support policy. Scientific research always is very indefinite. To reduce the degree of indefiniteness, special rules for presenting proposals are worked out, and the proposals are considered by

experts.   The authors are interested in getting money for their risky and indefinite projects.   The heads of the foundation are interested in as reliable investment of capitals as possible.   The scientific circles are interested in large-scale support of research.

When working out their criteria and ways of action, the DMs must be able to explain their policy in a natural language understandable to the authors of proposals, personnel of the foundation, and experts.

4.   After the secondary school young persons want to enter a university. Their choice is the matter of concern to the parents. The education in the majority of Russian universities is free, but the entrance is not so simple — the entrants must pass exams that are sometimes very difficult. There is competition to the best universities such as the Moscow University or Physico-Technical University, and only the entrants getting the highest score are admitted. Most commonly, the young people would like to explain their choice to the parents for which purpose a verbal rule of decision making must be worked out.

5.   A family living in a large town wants to lease a country-house for summer.   Since the members of the family appreciate different aspects of summer recreation and have different preferences, they decide to work out unique requirements to the proposed options. To do so, they decide to discuss various characteristics (estimates by criteria) of the country-house — for example, distance from the town, size of the house, quality of soil, etc. — which must be somehow taken into account in the unique rule for estimating the options of summer vacations.   The members of a family coordinate their decisions in an understandable and mutually acceptable language.

Features of unstructured problems

The above examples allow us to formulate the general features of the class of problems under consideration.

(1) These are unique problems of choice in the sense that each problem is new to the DM or has new characteristics as compared to the previous problem of this kind.

(2) The estimates of alternatives are indefinite because of the lack of information by the time of decision making.

(3) The estimates of alternative are qualitative and most often formulated in a natural language. Problem formulation is extremely sensitive to the way of describing it — sometimes a few words changed in the problem description can change the attitude of DM or active group.

(4) A general estimate of alternatives can be obtained only from the subjective preferences of DM(s). The DM's intuition and conviction that events will take one or another direction underlie the decision rule which enables one to pass from individual estimates to general estimation of alternatives.

(5) Estimates of alternatives by individual criteria can be obtained only from experts. An objective scale for measuring estimates by individual criteria is usually lacking. Moreover, the estimates of alternatives by criteria can be sometimes relative and show where one option is better than another.

Along with revealing the features of problems, the above examples enable us to judge about the DM's behavior in the process of decision making.

1. Being interested in the result, the DMs desire to control the entire process of choice. First of all, they seek that the experts not only estimate alternatives, but highlight those of their advantages and disadvantages that they would like to take into consideration in decision making. Additionally, any transformations of the original information, including their own information, they check for correspondence to their preferences.

2. The DMs prefer to communicate with representatives of active groups and other people in understandable terms of alternatives and preferences which enable them to work out compromise decisions.

The role of decision making methods

Let us ask: "What is the role of decision methods in unstructured problems?" These methods must help the DMs

(1) to structure their problems in collaboration with consultants,

(2) to work out a consistent policy for the problem of choice, and

(3) to consider complicated multicriteria estimates of alternatives.

Requirements on the methods of decision making

The features of the class of problems under consideration allow us to define the requirements on the methods for their solution (Larichev, 1987, 1992).

1. The methods must be adjusted to the language of problem description that is natural to the DMs and their environment. To be socially acceptable, the decision method must be readily adjustable to the accepted way of discussing problems in a particular organization. This usually means that the estimates of criteria and, consequently, the estimates of variants by the criteria are presented in a verbal form. Examples of descriptions of diverse problems can be found below in Appendix 1. They show that the verbal estimates on the ordinal scales of criteria make an adequate language for describing unstructured problems.

The decision method must be adjusted to such a description. Hence, by defining one or another form of verbal scales the DMs define the "measurer" for the experts estimating alternatives on these scales. The same verbal estimates are used by the DMs to define the requirements to alternatives, that is, the decision rule.

2. Since the construction of a subjective DM's decision rule is pivotal to the decision method, special attention must be paid to the capabilities and limitations of the human system of data processing. We believe that all operations of eliciting information from man must be carefully checked and for each of them the human "capacity limit" established. The operations undergoing such checking can be regarded as sufficiently reliable (see below). The use of well-checked methods of information elicitation provides a certain warranty of quality of the multicriteria decision methods. Without this warranty, one often cannot understand the quality of information elicited from DMs and experts.

3. If the criteria estimates are presented in verbal terms, then logic procedures can be applied to transform them while constructing the DM's

decision rules.   Logic transformations of verbal variables must be mathematically valid and based on verifiable conditions from which one or another form of decision rule follows.

4. The decision methods must incorporate means for checking the DM's information for consistency.   No matter what method is used to elicit information from DMs, one must be aware of the possibility of occasional errors and of the stages of DM training.   In this connection, procedures for checking the elicited information for consistency are required, as well as methods for detecting and eliminating contradictions in the DM's information.

The need for consistency checks is not eliminated by psychologically correct methods of information elicitation from DMs.   This checking is extremely important because it improves the efficiency of training and compels the DMs to recognize their errors and work out a reasonable compromise.

5. The decision method must be "transparent" to DMs, that is, they must have a possibility to check whether there is a correspondence between the resulting estimates of alternatives, on the one hand, and the estimates of experts and their own preferences, on the other hand.   This check allows the DMs to make sure that it is precisely their preferences that uniquely define the results of using this method.   Consequently, the DM must get explanations from the method in an understandable language.

Let us consider the above requirements in more detail.

Measurements

Requirement 1 refers to measurements carried out when analyzing unstructured problems. We are interested in measurements carried out before decision making, that is, in measuring factors that will be used in the further decision making. These measurements can be done by DMs themselves, but most often they are made by experts. Much depends on the results of these measurements. That is why it is vital to get a good (correct, regular, adequate) way of measurement. Obviously, it is both measurements and processing of their results that count here.

Quantitative measurements.

Thus, one could ask: "What about measurements in the area of decision making?" Let us turn to the classic works of J. von Neumann and O. Morgenstern (1947). According to their approach, to estimate alternatives and make decisions, one needs to measure the utility of alternatives. It is suggested to measure it in quantitative terms, though the human capabilities to do that have been somewhat doubted. J. von Neumann and O.Morgenstern were fully aware of the fact that the utility as a quantitatively measurable value could well be questioned. They understood that 'a direct sense of preference of some object or a set thereof as compared to others provides a necessary foundation.' However, they introduced an assumption, fundamental for their theory, that "our individual can compare not only events but also combinations of events with assigned probabilities." What was behind this assumption? It was based primarily on an analogy with physical measurements (of heat, length, light, etc.) where at the initial stages of human activity no methods of quantitative measurement existed, but emerged later.

In this connection, it was concluded that "Even if utilities look rather nonquantitative today, the history and case of heat theory may repeat again, and with unpredictable consequences at that" (von Neumann and Morgenstern, 1947). The evolution of the decision theory over the 35 years that elapsed since the publication of "Theory of Games and Economic Behavior" indicates that the expectations of its authors were over-optimistic. As it was the case in the past, we still lack reliable methods of quantitative measurement of probabilities and utilities. What is more, we have much evidence of unreliability of the existing methods for quantitative measurement of utility and probability.

There are two reasons for this phenomenon. First of all, many factors characteristic of decision making are of quite a different nature if compared with physical values. Factors, such as "organization's prestige", "originality of fashion", or "job attractiveness" are of extremely subjective character, and the personal values of an assessor affect considerable the measurements. Thus, there is no general objective basis with a chance of devising an objective measurement technique.

It is hard to imagine that any objective quantitative meters like thermometers, common for all people, ever will be available for such notions, since one can measure it only through peoples (just imagine that we would like to objectively measure quality of paintings or symphonies). In summing up numerous psychological research, we are safe in saying that there are no reliable methods of quantitative construction of the utility function in unstructured problems.

Let us consider in more detail to what extent the continuous quantitative scales are advisable as the language of measurement.

The estimates on criteria scales must reflect changes in value (utility, preference, importance, distinctness, etc.) of an alternative with a corresponding change of estimate by a given criterion. People are known to poorly estimate and compare objects of close utilities. With continuous scales, slight distinctions in estimates can result in different comparisons of alternatives. Indeed, all other estimates being equal, the preferability of an alternative will be defined by one insignificant difference.

The experiment of A. Tversky (Tversky, 1969) that demonstrated the stable intransitivity of choice was based precisely on this property. The subjects were given successive pairs of alternatives where gain slightly increased with a slight increase in payment. And they persistently preferred to get a higher gain for a slight increase in payment. Yet, when given the alternatives from the first and last pairs, they persistently preferred the first one, because they could not admit such a great increase in payment even for a corresponding increase in gain. In our view, the continuous scale of estimates prevented the subjects from seeing the gradual transition from quantity to quality. If the same subjects were given the same task but with ordinal (qualitative or verbal) estimates of payment and gain, then a transitive relationship could easily resulted.

Some our experimental results with choosing a summer country-house (Nikiforov et al., 1982) are indicative of the inconvenience of the continuous scales. Two continuous-scale criteria, cost and size of the territory, were used in these experiments. It was noted that for insignificant discrepancies in evaluation of the country-house the subjects do not necessarily find the alternatives that dominate the remaining ones. The subjects sometimes eliminated from the subset of the best alternatives (even if their number was only between four to seven) an alternative dominating one of the remaining alternatives. This observation can also be attributed to the fact that insignificant (5-7%) variations in cost do not affect appreciably the values of alternatives. Though the subjects assert that 'the cheaper, the better' if this difference is pointed out, on the whole they agree that both variants have the same utility.

The accuracy of human measurements on continuous scales was studied in experiments with three different decision support systems (Larichev et al., 1995b) applied to a typical unstructured problem of choosing a job. Two of the decision support systems (DSS) are based on the approach of the utility theory and differed only in minor details of measuring the weights of criteria and estimates of alternatives by the criteria. The third system ZAPROS (see the next chapter) is based on qualitative verbal estimates on the criteria scales.

The experiments demonstrated that the alternative rankings as established by the first two DSSs much less coincided with each other than with the results of ZAPROS. We note that the utility-based systems ranked all alternatives, whereas the ZAPROS DSS only established a quasi-order (with incomparable alternatives) which proved to be the most stable ranking. The relations between alternatives were close to dominance — the scales of alternatives were of ordinal nature.

The experiments suggest that the quantitative measurements are most sensitive to small errors and differences in the DMs answers, which gives rise to the question of the accuracy of human measurements, especially under indefiniteness. It is well known that in physics the accuracy of measurements depends on the precision of instrumentation. The same applies to human measurements. The available results of experiments are indicative of the fact that man cannot make precise quantitative measurements.

Transition from qualitative notions to numbers

A more reliable method of measuring many qualitative variables is urgently required because of the specificity of unstructured problems of decision making. This gave rise to many palliative approaches attempting to combine qualitative measurement scales and quantitative representation of the results. First of all, we should mention the simple means of establishing mutual correspondence between the primary qualitative measurement scale and the quantitative scale of scores where the primary measurements are carried out in qualitative form and (independently of the expert's will) are assigned certain numbers which are then used to estimate the variants of decisions. This method of measurement is not reliable because no logical basis underlies the assignment of one or another numerical value to the primary estimates. The worst of it is that the number are further treated as the results of objective physical measurements. For example, when estimating the quality of objects by multiple criteria, the scores by criteria are regarded as the results of quantitative measurements and are often multiplied by the weights of criteria and summed up.

When considering the problems of political choice, Dror (Dror, 1989) drew attention to the fact that people assign different numerical estimates to the same verbal definitions. We do not think that this necessarily means that one person believes that this event occurs with 70% probability and that 70% refers to a highly probable event, whereas another person believes that this event occurs with 90% probability and that 90% refers to a highly probable event. Both experts are, possibly, sure that this event is `very probable,' but when they are asked to evaluate this probability numerically (for example, in terms of percent or somehow else) they replace their ignorance of this number by some (rather arbitrary) number. Human estimates corresponding to the same verbal definition on a scale were experimentally shown to have a rather great dispersion (Wallsten et al., 1993) which is especially great for estimates representing the mean `neutral' level of quality.

The second popular approach is that of the theory of fuzzy sets where measurements are carried out in terms of descriptive qualitative values which are then transformed to the quantitative form by means of a given membership function assigning numbers to any word.

To what extent is this transformation reliable? To what extent is man error-free? It is obvious that the person constructing the membership function performs approximately the same operation as when establishing the correspondence between the qualitative and quantitative scales where the DM cannot evaluate the effect of small deviations in estimates on the resulting comparison of alternatives. The references to the check for sensitivity after quantitative measurements are of no avail. Indeed, in the presence of multiple quantitative parameters the sensitivity check becomes an independent involved problem which can be solved only by eliciting from the DMs information which they hardly can provide (see below).

The fuzzy-set approach of L. Zadeh is now a recognized lead of research in the applied mathematics. It enabled one to obtain many useful results for poorly structured problems having an objective basis for measurements. Yet, in the unstructured problems the accuracy of measurements based on the membership functions is doubtful.

Sensitivity check

It is believed by some researchers that accuracy of measurements is inessential in decision making — inaccurate measurements are admissible because analysis is completed by checking the result for sensitivity to variations of parameters.

Let us consider the ordering of multicriteria alternatives for the case where their utility is determined through the formula of weighted sum of criteria estimates. If, instead of a single estimate, a range is assigned to each numerical estimate of criterion weight and numerical estimate on scale, then the resulting utility estimate of each alternative also will lie within some range.

The ranges of different alternatives are, obviously, intersecting. How can one determine the rank of an alternative under inaccurate measurements? If alternatives with strongly intersecting values of utilities are declared equivalent, then the following can result: alternative A is equivalent to alternative B, alternative B is equivalent to alternative C, etc. If we make a transitive (in indifference) closure, then this often amounts to stating that all alternatives are equivalent, which makes the problem of ranking devoid of any sense.

Many questions about sensitivity arise when isolating the best alternative in a group. If there are only two criteria, then the problem is still simple enough; but if there are about five to seven criteria which is the case in many applications, then the sensitivity check for weights and criteria-based estimates becomes an independent and complicated problem.

Let us ask: "What is the aim of sensitivity analysis ?" If with variations in weights and estimates several alternatives by turns become the best, what can the analyst tell to the DM? It is difficult to convince the DM that the Pareto-optimal alternatives are close in utilities (including the alternatives with poor estimates by criteria that are important for the DM). The need for reliable measurements brings us to the need for of qualitative measurements in a natural language.

Verbal and quantitative measurements of event probabilities .

The verbal and quantitative measurements of probabilities can be compared using experimental data. As was established from experiments with many (442) respondents, 65% of them prefer to estimate probabilities in verbal form, and

70% prefer to get this information in quantitative form (Wallsten et al., 1993). This phenomenon was named the paradox of communication means (CMP). Let us consider an illustrative example (Erev, Cohen, 1990): there are three officers of different levels in an intelligence service. According to the CMP phenomenon, the middle-echelon officer would like to get from his subordinates quantitative information, but to report to his superior in verbal form.

Let us consider real-life examples. "Izvestya" published in 1995 a forecast of national economic development in 1996-1997 prepared by Aleksandr Lifshits, the then-councilor of President Yeltsin. The majority of his estimates were verbal. In 1981, the USA National Security Council prepared for President Reagan a list of five possible actions against Libya with verbal estimates of the chances for success (Hamm, 1991).

It might be well to note that the DMs' behavior is defined by their position in the organization, the higher-placed DMs require information in form convenient for analysis. Some facts cast doubts on the universality of the CMP phenomenon. For example, the former Secretary of State Henry Kissinger and President Gerald Ford favored the results of analysis (quantitative as well) if presented in verbal form. It seems that this phenomenon is characteristic of managers of middle hierarchical level.

There are some reasons why people prefer verbal information.

1. The verbal means of communication are more familiar and simpler, than the quantitative ones. We recall that the probability theory appeared much later than the languages of communication.

2. Since words are much more flexible and much less accurate means of expressing estimates than numbers (Teigen, 1988), they enable one to describe variables that are defined insufficiently.

On the other hand, there are reasons for preferring quantitative information.
1. Quantitative information is regarded as more accurate and reliable.
2. Quantitative information allows one to use many quantitative methods of transformation such as the Bayes theorem.

The experiments of Prof. T. Wallsten and his collaborators demonstrated that the form of probability estimates (verbal or quantitative) weakly influences the profit of subjects in experimental games (Budescu, Wallsten, 1995). We will describe in more detail one of the experiments (Erev, Cohen, 1990) which consisted of two stages where subjects played the roles of experts and DMs. The experts estimated the probabilities of various events in oncoming basket-ball matches, and the DMs ranked the events using their estimates. The experts estimated the event probabilities in both verbal and quantitative forms, and the DMs could choose that best suited to them. After the competitions, the DMs (and experts) received gratuity for correct answers. It turned out that the form of information exercised weak influence on the differences in gratuities which amounted to only 1 to 4%. We note that the . was defined in these experiments by one label, rather than by an evolved descriptive definition. One of the experimenters (Hamm, 1991) concluded that the convenience of verbal estimates is more essential than the financial losses.

Other experiments demonstrated that quantitative probability estimates can result in lower gains and also in dramatic increase in the so-called dominance rule (Erev et al., 1993). The experiments of Tversky (Tversky et al., 1988) demonstrated the intransitivity of preferences which is indicative of contradictory human behavior. It turned out (Gonzalez-Vallejo, Wallsten, 1992) that preference intransitivity is a much rarer occasion if verbal probability estimates are used. It was also established that if discussions are based on verbal probability estimates, then people reach agreement much faster (Erev et al., 1993), because somewhat different ideas are often implied by the same words, which facilitates coordinated group decision making (Gonzalez-Vallejo, Wallsten, 1992).

An analysis of former political decisions based on available documents (Gallhofer, Saris, 1989) presents an example where a certain decision is explained by some criteria of which one criterion is the probability of certain events with verbal scale of estimates. Why is this scale used? Because reasons and arguments in real decision making are presented in verbal form, which is human. This work not only demonstrated that people usually make use of these estimates, but also that a decision can be explained by choosing among alternatives with verbal estimates on criteria scales (without passing to numbers) and that the event probability is one of these criteria.

Comparative verbal probabilities

Some experiments focus on the relationship between the form of communication and degree of indefiniteness of events (Erev, Cohen, 1990). For example, the subjects were asked to estimate the chances of basket-ball teams to win in games between them. The experimenters noticed that in case of unknown team (higher indefiniteness) the experts were able to discriminate only two levels of verbal probabilities in comparative forms — for example, 'it is believed that the host always play better than the guests'. It is stated (Erev, Cohen, 1990) that compelling people to quantitative probability estimates in situations where only a few levels of indefiniteness can be discriminated can result in erroneous estimates. This example shows that some measurements can be carried out only in verbal form with the use of 'more probable than' relationships.

Methodical studies of comparative probabilistic estimates (Huber and Huber, 1987) demonstrated that comparative probabilities are much more frequently used by the common people (both adults and children) than quantitative estimates of probabilities of events. The experiments used tasks such as estimation of the probabilities of hitting the sectors of a rotating disk and estimating the winners in competitions and games. The authors of this work formulated six mathematical principles for comparative probabilities in form of axioms representing the mathematical concept of qualitative probabilities. We cite two of the axioms.

Let A, B, and C be events. For example, A is getting 5 at throwing dice, B is getting 2, etc. Let > denote 'more probable than'.

Principle 1: Transitivity.
If A>B and B>C, then A>C.

Principle 4:
If X is greater of or equal to A, then some event X is not less probable than the event A.

The main experimental result obtained with adults and children above five is as follows: the human comparisons follow completely the principles of the mathematical theory of qualitative probabilities. The authors of this work conclude that the six principles provide a more reliable foundation for describing human behavior than the laws of quantitative probability.

Instrumental analogy

Imagine that you have N unreliable balances weighing with great and poorly predictable errors. For example, you put the same object on balance several times and each time get different readings, that is, you cannot know which reading is true.

Let there be another, (N+1)st balance also measuring inaccurately. We weigh on it the N balances as objects. Now, let us assume that we have K objects and want to 'evaluate' them by weighing each of them on the balances, multiplying each measurement by the weight of the *i*th balances, and summing the results. The sum of N weights for us is the 'value' of each object. By comparing these 'values' we determine the object which is most valuable to us. Obviously, for balances with substantial errors the results of comparisons can be valid only if one object weighs on each balance much more than another. In all other cases the results of such weighing are questionable.

Unfortunately, we cannot improve our balances, but can 'calibrate' them before weighing by means of precise balance weights which are weighed on the imperfect balances, the results being laid off on a scale. Thus, the balances are 'marked'. One can readily see that the results of measurements based on this scale are incontrovertible only if one object is much heavier than another. In all other cases the results are questionable.

Now, we can give a descriptive explanation of the above example with balances. The inaccurate balances (measurers) are person(s) evaluating objects by N criteria and establishing the importance indices ('weights') of these criteria. It will be recalled once more that it is measurement of purely subjective factors in unstructured problems that is concerned.

There exist numerous experimental proofs of the inaccuracy of human measurements of subjective factors on continuous scales. We only mention the inaccuracies in measuring subjective probabilities (Kaheman et al., 1982) and in assigning weights of criteria (Borcherding et al., 1994).

The above measurements with 'calibration,' in our view, correspond to the transition from qualitative scales to quantitative ones by means of only the so-called membership functions or by establishing a direct correspondence between the qualitative notions and numbers.

Qualitative measurements.

We regard decision making in unstructured problems as the domain of human activity where quantitative (the more so, objective) means of measurement are not developed, and it is unlikely that they will appear in future. Therefore, it is required to estimate the possibility of doing reliable qualitative measurements. Following R. Carnap, we turn to the methods for measuring physical magnitudes that were used before the advent of reliable quantitative measurements. Before the invention of balances, for example, objects were compared in weight using two relationships — equivalence (E) and superiority (L), that is, people determined whether the objects are equal in weight or one is heavier than the other. There are four conditions to be satisfied by E and L (Carnap, 1969):

1. E is the equivalence relationship,
2. E and L must be mutually exclusive,
3. L is transitive, and
4. For two objects a and b either (i) a E b, or (ii) a L b, or (iii) b L a.

One can easily see that the above scheme enables one to carry out relatively simple comparisons of objects in one quality (weight). It is required here that all objects be accessible to the measurement maker (expert).

Let us consider another example, measurement of temperature. By applying palm to objects, people performed relative measurements based on the binary relationships E and L. At the next step, however, it was required to compare the measurements made by different persons (experts, so to speak) at different times or made by one person for different sets of objects. This became possible when people agreed about common points on the scales of measurement. For temperature, for instance, the following points could be defined:

(1) So hot that one can hardly apply the palm.
(2) The difference in temperatures (as compared with the body temperature) is almost not sensible.
(3) So cold that the hand freezes immediately.

We see that these definitions are not precise, but anyway they provide a ground for agreement. Using such definition, we have an absolutely ordinal scale with discrete estimates. Measurement is reduced to classifying objects as belonging to one of the estimates or to an interval between estimates.

Two more remarks are due. It is obvious that the thus-constructed absolute ordinal scale cannot have many values; otherwise, they will be poorly distinguishable by the measurement makers. To come to terms easier, it is required to identify commonly understandable and identically perceived points on the scale and explain their meaning in detail. Therefore, these scales must have detailed verbal definitions of estimates (grades of quality). Moreover, these definitions focus on those estimates on the measurement scale that were emphasized by the persons constructing the scale (for example, they could be interested only in very hot and very cold objects). Thus, the estimates on the ordinal scale are defined both by the persons interested in one or another kind of measurements (in our case, it is the DM) and by the distinguishability of estimates, that is, the possibility of describing them verbally in a form understandable to experts and DMs.

There is no reason to question the fact that before the coming of reliable methods of quantitative measurement of physical magnitudes, they were already measured qualitatively. These methods today could seem primitive because we have much more reliable quantitative methods. Yet, there is no doubt that the pre-quantitative (qualitative) methods of measuring physical magnitudes did exist. When they were superseded by by the quantitative methods, they were treated with negligence as something 'unscientific' and obsolete. The progress of physics gave rise to the well-known statement that the science appears wherever the number (quantity) occurs. To our mind, these declarations refer mostly to the natural sciences, but in the sciences dealing with human behavior qualitative measurements were and will be most reliable.

Final remarks

Let us sum up some results. There are multiple factors influencing estimation and choice of alternatives in human decision making for unstructured problems. Each of them has its own (mostly qualitative) language of measurement, although there are factors such as cost where a quantitative language would be quite adequate.

The accuracy (reliability) of measurements in decision making is extremely important because it defines the choice of the best alternative. It is only natural that the decision makers on whom human fortune depend want to put questions in a language understandable to them and get unambiguous answers. This suggest that for the majority of factors there exists a single possible kind of measurements — measurements in verbal notions arranged on ordinal scales.

Who must list the factors and estimate scales for them? For personal measurements, it is the person who not only indicated the factors to be considered, but the language of measurement as well, that is, the distinguishable levels of factors. For business decisions, sensible DMs consider not only their preferences, but also what could affect their decisions. If they turn to experts, they must clearly define, themselves or with somebody's help, verbal scales of estimates for the criteria. For qualitative factors, it is required to formulate in a natural language the grades of quality ordered on ordinal scale. Notably, this measurement language can be used to describe even very emotional problems of decision making. The reader is referred to Appendix 1 for the examples of measurement scales for three problems of decision making.

Thus, the method of verbal description satisfies the above requirement 1.

Valid information elicitation for constructing a decision rule.

Now, we turn to requirement 2 of psychologically correct methods of constructing the DM decision rules. We shall differentiate between two types of measurements. We discussed above measurements of main factors influencing decision. We shall refer to them as primary measurements. In some normative methods, primary measurements suffice for reaching the final decision. In the method of subjective expected utility, for example, the quantitative measurement of utility and subjective probability allow one to calculate the expected utility of every alternative.

Yet, for a large majority of normative methods this is insufficient, and some cognitive operations of information elicitation are needed to construct a decision rule. We will call them the secondary measurements. For example, one needs to measure weights of criteria to decide whether the utility function is additive or multiplicative (Keeney, Raiffa, 1976).

There exist many different normative methods of decision making which belong to the groups mentioned in Chapter 1. Their majority addresses problems with objective mathematical model or reliable scales for quantitative measurements. And only a small portion is intended for unstructured problems.

Normative decision methods present quite different requirements to their users such as 'assign weights to criteria,' 'construct the probability distribution of this outcome,' etc. To meet these requirements, an individual performs various operations of information processing which can be composite (incorporate other operations) or simple (elementary) operations that are not decomposable into elementary ones. Take, for example, the problem of constructing the utility function by a single criterion exercised within the framework of multiattribute utility theory (MAUT) (Keeney R.,Raiffa H.,1976). It involves a number of similar problems of finding a certainty equivalent for lotteries. The probability distributions are constructed on the base of such operations.

Analysis of different normative techniques described in Chapter 1, enables one to distinguish three groups of information processing operations such as operations with criteria, operations with estimates of alternatives by criteria, and operations with alternatives. Let us refer to an operation as elementary if it is not decomposable into simpler operations over to the objects of the same group, that is, to criteria, alternatives, and alternative estimates by criteria.

We propose to collect the results of psychological studies of the degree of human confidence and reliability in exercising one or another operation of information processing. If the data can be collected, then the psychological validity of a normative technique can be characterized in terms of psychological validity of the constituent elementary operations of information processing. We define the elementary operations as

- complex (C), if the psychological studies show that in performing such operations the decision maker displays many inconsistencies and makes use of simplified strategies, that is, eliminates a number of criteria,

-

- admissible (A), if the psychological studies show that the decision maker is capable of performing them with small inconsistencies and using complex strategies, that is, combinations of criteria estimates,

-

- admissible for small size (ASZ), if there are facts testifying that operations are performed rather reliably in case of few objects (criteria, outcomes, alternatives, and multicriteria estimates), but become increasingly more difficult as their number increases,
-
- uncertain (U, UC, UA), if the psychological research of these operations is insufficient. Yet, a tentative conclusion on admissibility (UA) or complexity of the operation (UC) can be drawn by reasoning by analogy from already known facts.
-

Table 2.1 contains the description of elementary operations and their estimates.

Table 2.1: The three groups of elementary operations and their estimates

No. of operation	Name of elementary operation	Evaluation
01	OPERATIONS WITH CRITERIA AS ITEMS	
011	Ordering in utility (value)	A
012	Assigning quantitative criteria weights	C
013	Decomposition of complex criterion into simpler ones	ASZ
02	OPERATIONS WITH SEPARATE ALTERNATIVE ASSESSMENTS BY CRITERIA	
021	Assigning a quantitative equivalent to qualitative estimate by a criterion	UC
022	Determination of quantitative equivalent of a lottery	C
023	Qualitative comparison of two estimates taken from two criteria scales	A
024	Determination of quantitative tradeoff value for two criteria estimates	UC
025	Determination of a satisfactory level by one criterion	UA
026	Nomination of probability for criteria estimate	C
03	OPERATIONS WITH ALTERNATIVES AS ITEMS	
031	Comparison of two alternatives viewed as a set of estimates by criteria and selection of the best one	ASZ
032	Comparison of two alternatives viewed as a whole, and selection of the best one	UA
033	Nomination of probabilistic estimates of alternatives	C
034	Attribution of alternatives to decision classes	ASZ
035	Quantitative estimation of utility	C
036	Decomposition of complex alternatives into simple ones	ASZ
037	Qualitative comparison of the probabilities of two alternatives	A

Let us describe each elementary operation in more detail.

Operations with criteria as items.

Operations 011 and 012 are measurements of comparative importance of criteria to the DM. Operation 011 is studied insufficiently, although there are some publications on subject's consistency in ranking of criteria. In case of seven criteria with binary estimates, the subjects were shown to rank criteria rather stably and consistently (Larichev et al., 1980). The subjects rank consistently criteria which are most important to them, although they tolerate permutations in the ranks of secondary criteria (Nikiforov et al,1984). Several recent papers (Fischer, 1991; Weber et al., 1988; Borcherding et al., 1993) showing that the subjects make substantial errors in quantitative measurements of criteria importance. Indeed, the quantitative measurement of weights is an unusual operation for a person hardly realizing its consequences (e.g., whether an insignificant variation in the weight of a criterion can result in choosing another alternative). Although this operation is used by many normative techniques (and is sometimes regarded as 'natural'), recent researches (Borcherding et al., 1993) show that weights assigned by the subjects cannot be regarded as reliable and stable information.

Operation 013 was studied when constructing the criteria hierarchy frequently employed in MAUT. The results indicate (Borcherding and Winterfeldt, 1988; Weber et al., 1988) that decomposition is not stable to DM's errors if the number of criteria is considerable. At the same time, it is our belief that a complex criterion can be decomposed quite reliably into two or three subcriteria that are obvious in terms of their meaning.

Operations with alternative estimation by criteria .

In our opinion, operation 021 is a groundless assignment of arbitrary numbers to the qualitative notions on scales. This operation seems difficult for the decision maker. Just as in assigning quantitative weights, an insignificant variation in numbers can affect the relationship between alternatives.

The reliability of preference measurement by lotteries was studied in detail by (Dolbear and Lave, 1967) who obtained negative results.

Operation 023 was studied methodically while developing the ZAPROS method. The results indicate that the DM performs it stably and with a small number of inconsistencies (Larichev et al., 1979).

Operation 024, that is, determination of the quantitative change in the estimate of one criterion that is equivalent to a change in the estimate of another criterion. Unfortunately, no methodic verification of reliability of this operation was carried out. In our judgement, here again we deal with measurements that are uncustomary to people.

As a number of studies show, operation 025 is a routine human operation of translating criteria into constraints. It is typically exercised while seeking admissible values and admissible (but not optimal) decisions (Kaheman et al., 1982).

Operations with alternatives .

Operation 031 was studied methodically by Russo (Russo and Rosen, 1975) who proved that it is difficult for humans. Even with four criteria, people use simplifying heuristics which may lead to errors and inconsistencies. It might be well to recall that it is precisely the pairwise comparisons of criteria estimates that allowed A. Tversky to build his `money pump' (Tversky, 1969), that is, the transitive ring of alternatives. At the same time, for two or three criteria this operation does not overstrain the short-term memory, which allows one to expect rather stable comparisons. Yet, the problem of pairwise comparison of two integral images of alternatives in form of photographs, slides, in some engineering tasks (Statnikov,Matusov,1995) etc., is regarded differently. We think that the decision maker develops a gestalt of alternative, which substantially increases the reliability of such comparisons. Unfortunately, we are unfamiliar with any methodical reliability study of such comparisons.

Operation 033 requires probabilistic estimates from the DMs. Yet, there is a convincing experimental evidence of their unreliability (Kahneman et al., 1982).

Operation 034 is employed in direct classification where a person classifies an alternative (described as a combination of criteria estimates) with any of several ordered decision classes. We have expressly studied the human abilities to handle such problems (Larichev et al., 1988). Reliability and stability of such information were shown to be dependent on the number of criteria, the number of estimates on scales, and the number of criteria classes. The domain of values of parameters for which people can handle this problem was established. Beyond this domain, the human behavior changes dramatically and manifests a dramatic increase in the number of inconsistencies and emergence of simplifying heuristics.

In operation 035, the utility of alternatives is measured quantitatively. If alternatives have monetary values characterizing profitability of one or another operation, then the quantitative language of measurements is adequate and the resulting estimates can be regarded as reliable, provided that the experts are reliable. At the same time, the quantitative estimates of utilities of various alternatives (ecological projects, scientific research, etc.) are unreliable (Johnson, Schkade, 1985) due to the inadequate language of measurements.

Recent studies on decomposing a complex alternative into simpler ones (Borcherding K.,1988) suggest that operation 036 is reliable only if it performs decomposition into a small number of relatively evident events. There are no special means of methodical checking of the listing of events for completeness upon constructing the trees of decisions and failures. We note that there is a need for additional psychological research of this operation.

Operation 037 can be regarded as reliable (Huber, Huber, 1987) because both adults and children perform it rather with confidence.

The operations in Table 3.1 are elementary. They can underlie more complex operations, and the estimates of simpler operations can be carried over to more complex operations. For example, the choice of the best alternative from a group is representable as a set of pairwise comparisons of alternatives (031), that is, it is admissible for small dimension. The operation of identifying the most unsatisfactory estimate of an alternative by a criterion is representable as a combination of operations 023, that is, is admissible.

Having the estimates of elementary operations, we can characterize the psychological validity of one or another normative technique. This approach was used to analyze some methods of multicriteria linear programming (Larichev et al.,1987). We note that only seven of the twenty considered techniques used the admissible operations.

Let us sum up. There are two reasons why one or another elementary operation of information processing proves to be difficult for the man. The first reason, which was mentioned above, is due to inadequacy of measurements, that is, attempts to use numbers where the man confidently employs only a few qualitative categories. We stress again that there exist two reasons why assignment of numbers to verbal ordered estimates reduces the reliability of measurements. First of all, by assigning numbers (scores, percent, etc.) the man constructs a subjective quantitative scale which can never be precise. By rating a qualitative estimate as 0.6 (rather than 0.65), the man can influence the final relationship of alternatives, that is, the decision. Second, the man cannot foresee how the numbers will be used in future (and different normative methods use them differently). It is the inadequacy of measurements that makes the elementary operations 012, 021, 022, 024, 026, 033, and 035 complicated.

The second reason is directly related to the limitations of the short-term memory. Operations 031 and 034 require that the memory stores the criteria estimates of alternatives, as well as the classes of decisions (for 034). It is namely owing to the limited capacity of the short-term memory that for a small values of some parameters these operations are performed sufficiently reliably and unreliably for larger values.

Which elementary operations can then be considered as valid and admissible? They are far from many, and all of them, except for 025, are qualitative. We may somewhat reliably use operations 011, 023, 032, and 057. We may, within some limits, use operations 025, 031, 034, and 036. As for the admissible operations, they boil down to qualitative comparisons (of the type 'better,' 'worse,' or 'approximately equal') of criteria, pairs of estimates on two criteria scales, and holistic alternatives. We can also assign satisfactory values and perform a simple decomposition of criteria and alternatives. For a small number of criteria, we can compare two alternatives. For not-too-large number of criteria, decision classes, and grades on scales, we can classify alternatives into decision classes.

In totality, all this seems a significant restriction for the researcher working on normative methods. Yet, as will be shown below, it is possible to develop psychologically valid decision methods.

We conclude by discussing some important issues.

First, many of the above negative estimates of information processing operations are based on laboratory studies. One may anticipate that human behavior in this environment differs essentially from that in real life. Our studies (Larichev, Moshkovich, 1980) are also indicative of the existence of such differences. Yet, numerous studies of human behavior in real life such as making political decisions, races, gambling, etc., reveal that the typical characteristics of human behavior here also manifest themselves, though, possibly, in another form. Obviously, experiments are sometimes insufficient, but if their result is negative, then it is hardly advisable to make use of unreliable means when solving important problems.

Second, when speaking about human errors and contradictions we in no way imply that the man is an 'intellectual cripple.' On the contrary, the human system of information processing is perfectly suited to the majority of problems that challenged the humankind in the course of its development. Within certain limits, the man is suited to multifactor problems if the number of factors is small enough. Moreover, people have a set of heuristics enabling them to solve problems of any complexity by simplifying them and adjusting to their limited capabilities. Yet, there exist problems that are difficult for man. There is nothing amazing that they exist. In the final analysis, the humankind is a community of biological beings limited in all respects. Man cannot make a five-meter standing jump, or do without water for five days, etc. In just the same way, man cannot consider multiple factors without a recourse to heuristics. All heuristics are good in the majority of cases, but sometimes they result in inconsistencies and contradictions. As A. Tversky (Tversky, 1969) demonstrated convincingly, a simple device of neglecting minor differences in criteria between two alternatives leads to intransitivity.

So, the multifactor problems that made their appearance in the course of recent decades are especially difficult to people and fraught with danger of errors stemming from heuristic devices adjusting them to the limited capabilities of the human information processing system.

Third, one can conclude that the preferability of getting from people qualitative and not quantitative information is beyond question.

Valid transformation of verbal information

We turn to requirement 3. Appendix 2 to this chapter overviews briefly some of the existing methods for comparing alternatives in terms of qualitative information such as Pareto dominance, lexicographic ordering in importance of criteria, pairwise compensation of criteria by ordering them in importance, or construction of a unified order scale. The overview suggests that qualitative information and logic rules for using it do not eliminate the need for checking some conditions that are necessary to construct valid decision rules.

The existing rules of qualitative evaluation of multicriteria alternatives are based on preference independence of one or more criteria . In general, this property can be formulated as follows. Let there be a set of criteria K decomposed into f disjoint groups K_1, $K_2,,...,K_f$. Then we have the following definition.

Definition 2.1. The criteria of group K_m ($1 < m < f$) are preference independent of the rest of criteria of this set if the preference between alternatives having identical estimates by all criteria, except for those of the m-th group, is independent of particular values of these identical estimates.

The preference independence conditions were introduced in (Keeney, Raiffa, 1976) where they underlie theorems on the properties of the utility function for quantitative measurements. The results obtained there are universal and can be applied to qualitative measurements as well. The use of the preference independence conditions stems from the desire to construct an efficient (in the sense of high degree of comparability of real alternatives) decision rule from relatively scanty information about the DM preferences. On the other hand, an exhaustive independence check involves comparison of very great (sometimes prohibitively great) number of vector estimates. Therefore, there exist two problems: (1) how to carry out a partial, but sufficiently representative check and (2) what should be done in case of dependent criteria? Let us discuss them one after the other.

Kinds of dependencies of preference criteria and methods of their checking

The following kinds of dependence of the preference criteria can be distinguished:

(1) dependence of a criterion on one or more criteria,
(2) dependence of a pair of criteria on one or more criteria, and
(3) dependence of a group of criteria on one or more criteria.

The dependence of one criterion on other criteria means that the condition of ordinality of criteria scales is violated. In a broader sense, this may imply that there exists a nominal criterion, that is, criterion whose scale defines for the DM some possible values which are not preference-ordered. For example, to the question "What do you like, tea or coffee?" an individual could answer "tea;" but if we add to the definition the time of day, then the answer could be "I like coffee in the morning and tea at night." Therefore, the preferability of estimates `tea' and `coffee' on the `beverage' scale depends on another criterion, `time of day.' It might be well to note that this form of dependence is often overlooked by the experts in decision making and the thus-constructed scales (separate for each criterion) are regarded as valid for any combination of estimates by other criteria. At the same time, it is very easy to check this condition, and the way out can be found in merging such pairs of criteria into one criterion with an ordinal scale of estimates.

The dependence of a pair of criteria on the rest of the criteria is the best-understood case. It might be well to note that this kind of (in)dependence of criteria is the mainland of the decision methods. As was proved, if all m criteria are pairwise independent, then there exists an additive value function and, as a result, any group of criteria is independent of the rest of criteria. We refer also to the opinion of D.vonWinterfeldt and R. Fischer (Winterfeldt, Fischer, 1975) that the group dependence of criteria "is indefinite in nature and difficult to detect" if the criteria are pairwise preference-independent. Therefore, we can understand the researchers trying to construct procedures for validation of the axiom of preference independence of pairs of criteria.

It is proposed (Gnedenko et al., 1986) to carry out this validation also on specially designed pairs of alternatives. Here, the DM is asked to compare all possible pairs of alternatives where all estimates but one are at the most preferable level. For each such pair, the estimates (by all criteria but those two where they differ) are then replaced by the least preferable ones and these pairs are presented to the DM for comparison. Thus, the check is carried out in a similar manner, but the DM needs not to indicate the indifference points, but has only to compare the alternatives differing in estimates by two criteria. In doing so, operation 031, which is admissible for two criteria, is used. If the pairwise comparison does not reveal dependence of criteria, then it is more probable than not that there is none.

This check, undoubtedly, is not exhaustive; but if one takes into consideration that the best fixed values in criteria are replaced by the worst ones, that is, the quality of the fixed 'residue' is reversed to a diametrally opposite value, then it seems that one can assert to a great degree of confidence that if here the DM's preferences are identical for all pairs of vector estimates, then the criteria are pairwise preference-independent for the DM.

Importantly, all above methods of validating the conditions of criteria independence can be realized in terms of verbal estimates. Therefore, the check for independence can be realized within the framework of the approach developed under this cover, verbal decision analysis .

Appendix 3 presents simplistic examples of dependence of two criteria on a third one for choosing

a summer country-house,
the camp in canoe tourism,
apartment to lease.

We note that in the examples of Appendix 3 the ordinality of scales of individual criteria is not violated by the existence of dependent criteria. Indeed, availability of a shop is always preferable to its absence (for any distance of the country-house from Moscow); a good bathing place is always preferable to a bad one (for any amount of mosquitoes); it is always preferable to reach job in twenty minutes to in one hour (independently of the availability of a local train). Therefore, the question here is only of a preferable set of estimates by different criteria.

Changes in problem description in case of dependent criteria.

What to do in case of dependent criteria? McCrimman (McCrimman, Wehrung, 1975) proposes to group together dependent criteria and reformulate them as a single criterion independent of the rest of them. The same is proposed by Kelly, the author of the theory of personal constructs (Kelly, 1955).

It is worth noting that in all of the above example we can group all three dependent criteria. However, it often happens that, in addition to its meaning in the dependent group, one of criteria has a complementary meaning. Appendix 3 shows how one can identify dependent criteria and transform the description of a problem in order to obtain a set of preference independent criteria.

Thus, in case of dependent criteria it is advisable to reformulate the problem by merging the dependent criteria into a single criterion and retaining that carrying independent complementary sense. This can be done conveniently by means of verbal estimates of alternatives.

Another method of eliminating the dependence of criteria is worth noting: the hierarchy of criteria if they make up an obvious group of indices characterizing some generalized quality of the object. In this case, it is possible to make up groups of independent criteria and within each group to order the combinations of all possible estimates by these criteria. If the number of combinations is large (more than 5--7), then they are decomposed into ordered groups representing estimates on the scale of a new generalized criterion. For example, up to fifteen criteria were used to assess programmers (Rajkovic, 1985). In doing so, the criteria such as `initiativity,' `creativity,' and `sociability' were naturally merged into generalized criterion of `personal characteristics' for which an ordinal three-valued scale (`very good,' `fair,' and `unsatisfactory') was used.    One of the above estimates by the generalized criterion `personal characteristics' was assigned to each possible combination of estimates by the first three criteria.

Therefore, we can conclude that in the decision methods based only on qualitative verbal estimates it is necessary:

(1) to check for pairwise preference independence of criteria by comparing alternatives differing in estimates in two criteria (admissible operation 023),

(2) identify independent groups of criteria,

(3) detect interrelationships of the dependent criteria and the nature of this dependence, and

(4) reformulate some criteria so that the resulting criterial description of the problem enables one to employ efficient methods of comparison of multicriteria alternatives without disregarding the indices which are important for decision making.

Thus, to satisfy the above requirement 3, one has to transform correctly the verbal problem description to a new description with independent criteria.

Now, we go to the next problem, that of detecting and eliminating errors in information elicited from people in the course of decision making.

Closed procedures

Now we turn to requirement 4. It has been known since the time of antiquity that "To err is human." People err when transmitting or processing information. They err less, maybe substantially less, if they use the above valid procedures of information elicitation, but, nevertheless, they do err. The errors can be due to distraction, fatigue, or other causes. They are observed in real life, as well as in psychological experiments. They differ essentially from the human errors in psychometric experiments which are known to follow the Gauss law and have greater probability for greater deviations from the true value. The human errors in procedures of information processing are of different nature. For example, our studies of multicriteria classification demonstrated that in problems of small dimension (which are simple for man) gross errors leading to many contradictions are rare — 1 or 2 out of 50 cases. These errors are obvious. Errors of the same kind are met when comparing pairs of estimates by criteria, ranking criteria, etc. Stated differently, man can once and again commit essential errors. Therefore, the information elicited from man must be validated and not used uncontrollably.

What methods of information validation are in existence?

The simplest method is to ask once more. Even the simplest question "Are you sure?" intensifies human attention in man-computer dialogue. On the other hand, if the question is repeated literally, then the man can mechanically give the same incorrect response. That is why in psychological experiments questions are repeated in queries at different points and spaced in time (e.g., the first question is repeated after the twentieth question, etc.). Clearly, the disadvantage of this method of monitoring is large number of questions.

Much more efficient means are offered by closed procedures where the previous information is validated indirectly, rather than directly. The questioning procedure is constructed so that questions are duplicated implicitly via other questions that are logically associated with them. Let us consider an example. We want to compare four values: A, B, C, and D. The easiest way of solving this problem is to perform a successive pairwise comparison. Let the comparison establish that $A > B$, $B > C$, and $C > D$, which allows us to order the values as $A > B > C > D$. Since there can be errors in the DM's responses, it seems logical to compare each value with the others. We note that the detection of inconsistency must stimulate logical analysis, rather than to eliminate automatically the error and averaging of answers.

So, what is the proper way of building closed procedures for psychologically valid operations of information processing. For any operations of comparison of several values, we may introduce additional objects of comparison. For operation 011, one can easily compare each criterion with every other criterion and analyze the inconsistencies emerging while ordering the criteria by this comparison. Operation 032 can be checked in a similar way.

The following closed procedure was first suggested for operation 023 in (Larichev et al.,1974). Let there be Q criteria with ordinal scales and a small (2-5) number of estimates. It is required to order the estimates of all criteria, that is, to arrange them on a joint ordinal scale. To this purpose, it was suggested to perform pairwise comparisons of criteria scales. .

All 0.5Q(Q-1) pairs of criteria were pairwise compared, which enabled a rather reliable validation of the DMs' information. We note that as the number of criteria (hence, the complexity of the problem) increases, the potential amount of redundant information generated by this comparison increases as well. A closed procedure of this type has been employed to advantage in the ZAPROS method (see Ch. 3).

The information elicited using operation 034 was validated by the Pareto-dominance relationship. For the alternative A classified by the DM with some class of decisions, in the space of criteria a cone of Pareto dominance was constructed, the alternatives getting into it must belong to a class not worse than A. If this condition is violated, then contradiction occurs which the DM is asked to analyze. This type of validation procedure was used in the ORCLASS method (see Ch. 5).

The information elicited using operation 031 can be validated in two ways. The first one is to use operation 023, that is, to compare the changes in estimates for criteria pairs, with the aim of establishing the preference of one alternative over another. We note that here alternatives are not always comparable and one of the the results can be incomparability.

The second approach can be used together with the first one if there are more than one alternative. Then one may try to construct a closed procedure by pairwise comparison of alternatives. It is sometimes advisable to introduce an additional alternative which is not related by dominance with the two main alternatives and which enables additional validation of the DM's preferences.

We do not know how to validate the decomposition operation 013. Therefore, we think it would be advisable to use it if the decomposition is obvious and the number of elements of the next level is small.

Learning procedures

As was noted above, learning is human. It is one of the inherent properties of human behavior, and the method of trial-and-error is the most characteristic human feature. Learning involves the study of a multicriteria problem and gradual working out of the DM's policy (decision rule).

As many researchers supposed, man has no preconceived decision rule. As noted in (Winterfeldt, Edwards, 1986, p. 351), it can be hardly expected that the utilities and numbers expressing the subjective estimates of objects and situations are just stored in our minds until elicited. Despite the fact that such expectations were not made explicitly, they were implied. Indeed, in many decision methods people are required to give immediately all parameters of decision rules. It can hardly be expected that at the initial stages of decision making an individual can define sensibly and consistently the decision rule. It can be assumed that an experienced DM (especially, that who dealt previously with such a problem) has some elements of policy such as a (possibly, incomplete) list of criteria, comparative importance of some criteria and estimates, etc., but usually all this is specified in the course of decision making where all tradeoffs are defined.

To allow the human ability of learning to manifest itself, the decision method must comprise special procedures for gradual, rather than instantaneous, working out of the DM's policy. These procedure must allow the individuals to err and correct themselves, to work out partial compromises, and go on to the next ones. This process must allow the individuals to challenge their own decisions and return to the beginning.

Such possibilities are offered by the new decision methods discussed in the chapters to follow. All of them include special procedures for working out the decision rule. By erring and correcting themselves, the individuals work out a consistent and well thought-out strategies. These procedures enable one to satisfy requirement 4.

Explanations

Now we turn to requirement 5. From a behavioral point of view, one of the requirements on any method is explainability of its results. The DM making a responsible decision would like to know why alternative A is superior to B and why both are superior to C. This requirement is quite legitimate. The stages of information elicitation from DM (measurements) and presentation of the final results are separated by information handling. Understandably, the DMs want to make sure that the assessments of alternatives are based, without any distortion, precisely on their own preferences. To meet this requirement, the decision method must be `transparent,' that is, allow one to find the one-to-one correspondence between the DM's information and the final evaluations of alternatives. Only then one can expect to get explanations from the DM.

The possibility of getting explanations in a natural language is one of the characteristic properties of expert systems which enables one to make the systems user-friendly. In problems of decision making, it is the DM who is a user and whose preferences underlie the decision rule. As will be seen below, for valid measurements of variables there exists a possibility of getting an explanation in form of "Such and such relationships hold on the basis of such and such information elicited from the DM and checked for consistency". These explanations enable one to satisfy the above requirement 5.

The new approach and construction of decision methods

How can the proposed approach be used in developing a new prescriptive decision method? First of all, we again emphasize that it must be oriented not to an unreflecting apparatus of information processing, but to a real person perceiving information in a certain form, limited in information-processing capabilities, and liable to errors and inconsistencies.

First of all, the source information for any method is a structured description of the problem in a natural language and in terms understandable both to DMs and would-be experts. The use of multiple criteria with verbal estimates on ordinal scales is the softest method of structuring which introduces no distortion. This description must be retained, without distorting it by transformations, at all stages of using the developed method.

Second, a prescriptive method must be oriented to particular problems. The most important characteristics of problems are form of solution (for example, determination of the best alternative, division into ordered classes, etc.) and information about the alternatives under consideration (for example, availability or absence of real alternatives at the time of decision making, the number of alternatives and criteria describing them, etc.).

Third, the possibility of analyzing the problem description for adequacy, completeness, etc. must be envisaged. The check of identified criteria for independence is an important point in such an analysis. If the criteria are found to be partially dependent, then it is advisable to restructure the problem description so that the new criteria (or their groups) be independent.

The requirements to the method together with the problem description provide the source information for choosing the admissible valid operations for elicitation of information which lead to the desired decision rule. Sometimes, this choice can be simple; sometimes the desired result can be reached only along difficult paths.

Fourth, the method must have means for checking information for consistency, no matter which technique is used to elicit it. A process of elicitation and validation of information must be built which facilitates the learning of necessary compromises.

Finally, the method must provide facilities for explaining decisions to the DMs in their natural language.

Appendix 1

Problem of planning research and development

Criteria

A. OUTLOOK of the project

(1) realization of the project can result in basically new theoretical concepts, experimental methods, industrial technologies and materials;

(2) realization of the project will result in a significant development of theoretical concepts, improvement of experimental methods, industrial technologies and materials

(3) realization of the project will result in some improvement of experimental methods, industrial technologies and materials.

B. *NOVELTY of the approach to formulated problems*

(1) original approach that was not met earlier,

(2) updating of the existing approaches,

(3) replication of the existing approaches.

C. *QUALIFICATION of project executors*

(1) the project executors are one of the best research teams in terms of experience and qualification;

(2) the level of experience and qualification of the executors is sufficient for the project;

(3) the executors lack the required experience and qualification.

D. *RESULTS of previous work on the problem*

(1) a significant progress was made within the framework of this problem;

(2) some progress was made;

(3) the progress is insignificant, and mostly preliminary research is being carried out.

E. *RESOURCE support of research*

(1) the executors have sufficient materials and equipment for completing successfully the project by the deadline;

(2) the executors have some materials and equipment, but additional resources are required to complete the project by the deadline;

(3) the executors are poorly equipped with resources required for successful work.

F. Possibility of fast UTILIZATION of the results

(1) the results will be sufficiently efficient to be rapidly introduced into practice;

(2) additional research and development will be required to introduce the planned results into practice;

(3) the work will be of largely theoretical nature.

Problem of choosing universities by entrants

Criteria

A. COMPETITION

(1) small competition — 1.5 to 3 persons per place;

(2) average competition — 4 to 5 persons per place;

(3) large competition — 8 to 10 persons per place.

B. COMPLEXITY OF ENTRANCE EXAMINATIONS

(1) examinations are not difficult, the level of school finals;

(2) examinations are more difficult than the school finals, but are sufficiently regular to allow one to read up for them;

(3) examinations are difficult and nonstandard.

C. UNIVERSITY PRESTIGIOUSNESS

(1) the university is regarded as one of the best in the country;

(2) the university is classified with high-level universities;

(3) the university is regarded to be of moderate level.

D. ATTRACTIVENESS OF THE PROFESSION given by the university

(1) you like very much the profession given by the university;

(2) you do not like very much the profession given by the university, but it is acceptable to you;

(3) you do not like the profession given by the university.

E. CORRESPONDENCE OF THE PROFESSION given by the university to your personal qualities and abilities

(1) the profession given by the university perfectly corresponds to your personal qualities, abilities, and dispositions;

(2) the profession given by the university does not correspond in all respects to your personal qualities, abilities, and dispositions, but is generally acceptable to you;

(3) the profession given by the university does not correspond by any means to your personal qualities, abilities, dispositions, etc.

Choosing the dress style

Critera

A. CORRESPONDENCE TO THE FASHION:

(1) all the vogue;

(2) fashionable dress;

(3) this fashion can be worn at any time;

(4) this fashion going out of mode;

(5) this fashion is obsolete.

B. ORIGINALITY OF THE DRESS

(1) original;

(2) not original.

C. CONVENIENT TO WEAR

(1) I will well in it;

(2) I will normal in it;

(3)I will be constrained in it.

D. MATCHING OF MATERIAL COLOR

(1) very good;

(2) a slightly different hue is required;

(3) another color is required.

E. ATTRACTIVENESS

(1) undoubtedly, I will like it;

(2) most likely, I will not like it.

F. TIME

(1) this fashion was and will be worn (independently of time);

(2) this fashion will be in fashion for a long time;

(3) this fashion will be in fashion for several seasons;

(4) this fashion will pass soon;

(5) this fashion will not be accepted.

G. SENSE OF MEASURE in this model

(1) there is sense of measure in the model;

(2) there is no sense of measure in the model.

Appendix 2

Let us consider some of the existing approaches to comparing and estimating alternatives on the basis of qualitative information. We assume that there are alternatives A and B estimated by Q criteria with ordinal scales: q is the number of estimates on the scale of q-th criterion, the estimates on the scale of each criterion are arranged in descending order of preference, and the alternatives A and B are represented by the vectors of estimates in criteria

$$A = (a_1, a_2, ..., a_Q), B = (b_1, b_2, ..., b_Q)$$

Dominance

Dominance, that is, componentwise comparison of alternatives (identification of the nondominated alternatives by the principle of Pareto), is the simplest rule for comparing multicriteria alternatives on the basis of qualitative information.

Definition 2.2. Alternative A dominates alternative B if its estimates in each criterion are not less preferable than those of B and at least in one criterion estimate of A is better.

This definition can be represented formally as

$A \succ B, if \ \forall i = 1,2,...,Q, a_i \geq b_i; \exists j \ such \ that \ a_j > b_j,$

where $\geq$ stands for 'not less preferable' and $>$ stands for 'more preferable'.

This popular rule is used in many decision methods upon initial contraction the original set of alternatives. It might be well to note that it is applicable if the condition of independence (in preference) of each criterion of the rest ones is met, which means that the preferability of estimates on the scale of this criterion is independent of the estimates by other criteria.

The disadvantage of this rule is that more often than not it does not provide a solution.

Importance of criteria.

The following examples of decision making from qualitative information are based on ordering in importance the criteria by which alternatives are estimated. Notably, although the notion of importance of criterion (or criteria) is used frequently, in essence no attempt was made to define and use it. The notion of relative importance of criterion was considered by Sayeki (Sayeki, 1972) and Podinovskii (Podinovskii, 1978). The latter author defined the preference (importance) of criteria as follows:

Definition 2.3. Criterion K_1 is preferable to (more important than) criterion K_2 if in all possible pairs of alternatives differing in these two criteria the alternative whith a more preferable estimate in K_1 is preferable.

This definition is valid for homogeneous criteria, that is, criteria having the same ordinal scales; but it can be extended to other criteria with ordinal scales distinguishable from the point of view of utility (value) for the DM.

We prove that the satisfaction of this condition implies that K_1 and K_2 are preference independent of the rest of criteria.

Indeed, the preference independence of K_1 and K_2 of the rest of criteria means that the comparison of any two alternatives differing only in estimates by these two criteria is independent of the values of equal components by other criteria. The above definition, however, indicates that preference is given in this case to the alternative with the most preferable estimate in criterion K_1. Consequently, the result of comparing such pairs of alternatives is independent of the values of equal components by other criteria.

It is possible to introduce a notion of additive importance of criteria which also is used, as will be seen in what follows, in the rules of comparison of preference alternatives.

Definition 2.4. Criterion K_1 is preferable to (more important than) criteria K_2 and K_3 taken together if in all possible pairs of alternatives differing only in estimates by these three criteria the alternative with a more preferable estimate in K_1. is preferable

By analogy with Definition 2.3., criteria K_1, K_2, and K_3 can be shown to be preference-independent of the rest of them. These notions undoubtedly are strong enough requirements in the system of DM preferences. On the other hand, they can be validated and used in rather popular methods of comparing multicriteria alternatives that are discussed below.

Lexicographic ordering.

The best-known rule of decision making by ranking criteria in importance is that of lexicographic ordering which first chooses the alternatives with the best estimates by the most important criterion, then chooses among them the alternatives with the best estimates by the second-in-importance criterion, and so on.

Definition 2.5. Alternative A is more preferable than alternative B if it is more preferable by a more important criterion.

Formally, this rule is representable as
$A \succ B, \text{if } a_i > b_i$
and K_i is the most important criterion or $a_j = b_j$
for all criteria K_j more important than K_i.

Obviously, this rule dwells on the assumption that the most important criterion is preferable to the rest of them taken together, the second-in-importance criterion is more important than the rest of them (except for the first one) taken together, etc.

The disadvantage of this rule is its noncompensatory nature, that is, the possibility of compensating the disadvantages of an alternative in some indices by 'good' estimates in other indices is disregarded; that is why this rule finds rare application in practice.

Order of pairwise compensation

Ladenzon and Litvak (1988) suggested a rule for comparing alternatives on the basis of a simple ordering of criteria in importance. It is based on pairwise compensation.

Definition 2.6. Alternative A is more preferable than alternative B if for each criterion by which A is less preferable than B there will be a more preferable criterion by which A is more preferable than B.

Formally, this rule is representable as
$A \succ B, if \ \forall i \ such \ that \ a_i < b_i, \exists j \ such \ that \ a_j > b_j$
and criterion K_j is more preferable (important) than K_i.

We note that this rule is valid if the importance of criteria is understood in the sense of the above definition. Therefore, the condition of preference-independence of a pair of given criteria of the rest of them must be satisfied.

Thus, this rule implies that the criteria are pairwise preference-independent.

An obvious disadvantage of this rule resides in that it disregards the difference in estimates by criteria, that is, the introduced notion of importance of criteria imposes very strict requirements which are not always satisfied in practice even if the DM easily orders the criteria in importance.

Joint ordinal scale (pairwise compensation)

This scale enables one to construct a partial ordering of alternatives with regard for the effect of differences of estimates by individual criteria. It is constructed by comparing special pairs of alternatives differing in estimates by two criteria

(Larichev et al., 1974). On this scale, the estimates by all criteria are ordered. The use of this scale for comparing real alternatives also is based on the idea of pairwise compensation.

Definition 2.7. Alternative A is more preferable than alternative B if for each estimate of A on the joint ordinal scale there will an estimate of B which is less preferable.

Formally, this rule is representable as

$$A \succ B, if\ \forall a_i (i = 1,2,...Q), \exists b_j, such\ that\ a_i \geq b_j\ and\ \exists k, such\ that\ a_k > b_k$$

It might be well to note that, for this rule to be valid, it is required that the criteria also must be preference independent . Therefore, one can conclude that there exists a set of rules for comparing alternatives that make use only of qualitative estimates of DMs and experts.

Appendix 3

Problem of choosing a summer country-house

Table 2.2. The alternatives of a summer country-house.

Alternatives	Criteria		
	Quality of country-house (comfort)	Availability of shop near	Distance to Moscow
A	good	no shop	
B	average	there is a shop	

It is quite possible that alternative A is preferable to B if their estimate by the last criterion is `near Moscow.' At the same time, if their estimate is `big distance to Moscow,' then B can prove to be preferable to A.

If you choose a country-house, then the lack of shop is inessential if it is located near Moscow, and essential, otherwise. At the same time, good and average quality of the country-house are of the same value to the DM, both for small and large distance from Moscow. Therefore, it suffices to merge the criteria `availability of shop' and `distance to Moscow' into a single criterion. Here, if other variants of country-houses are considered, then information about closeness to Moscow has sense both if there is shop or not.

Choice of a camping place

Table 2.3. The alternatives of camping place.

Alternatives	Criteria		
	availability of firewood	*quality of bathing place*	*availability of moscitoes*
A	much	bad	
B	little	good	

It is quite possible that alternative A is preferable to B if their estimate by the last criterion is `a lot of moscitoes.' At the same time, if their estimate is `no moscitoes,' then B can prove to be preferable to A.

If we consider the second example (choice of camping place), then the criterion `availability of moscitoes' influences the value of estimates by `availability of bathing place.' As above, here the criterion of `availability of moscitoes' has independent importance (if other places of camping are considered).

An apartment to lease

Table 2.4. The alternatives of an apartment to lease.

Alternatives	Criteria		
	habitation	*time to work*	*availability of local train*
A	house	one hour	
B	apartment	20 min	

It is quite possible that alternative A is preferable to B if their estimate by the last criterion is `there is local train.' At the same time, if their estimate is `no local train' then B can prove to be preferable to A.(Humphreys,1977).

In the last example, the criterion `availability of local train' changes the values of estimates by `time to work,' but does not affect the estimates by `habitation.' It seems to be of no importance for comparing other variants of habitation. The criterion `time to work' has here an independent importance.

We demonstrate now by way of examples how to change the description of a problem with the aim of transforming criteria into preference-independent ones.

In the first case, it is only logical to reformulate `availability of shop' and `distance to Moscow' into a single criterion of `possibility of getting food' with estimates

(1) good, if there is a shop or the country-house is near Moscow, and

(2) bad, if if there is no shop or the country-house is far from Moscow.

We now have a new independent criterion, but also retain `distance to Moscow' which carries an additional meaning.

In the second example, it is required to reformulate `quality of bathing place' and `availability of moscitoes' into `possibility of pleasant bathing' with estimates

(1) good (good bottom, convenient approach, no moscitoes),

(2) bad (bad bottom, inconvenient approach, moscitoes).

We replace `quality of bathing place' by this criterion and retain `availability of moscitoes' as having additional sense.

In the last example, we merge `time to work' and `availability of local train' into `possibility of convenient transportation' with estimates

(1) good (habitation near work or there is a local train),

(2) bad (habitation far from work or there is no local train).

Here, `time to work' is retained as having independent meaning.

3 THE METHOD **ZAPROS-LM** FOR PARTIAL RANK-ORDERING OF MULTIATTRIBUTE ALTERNATIVES

Main ideas of the method ZAPROS-LM

Let us illustrate the ideas of the proposed method on a simple example of the task of evaluating R&D projects submitted for research grants.

How to evaluate R&D projects

A group of sponsors decided to organize a Fund for investing money into applied R&D projects. It is known that such Funds exist in many countries and that in some area special scientific projects can lead to an essential financial success.

The Fund organizer was interested in developing an effective system for projects' selection. To carry out the job a decision analysis consultant was invited. It was decided that each project had to be prepared in accordance with the strict regulations, and that the projects' expert evaluation was to be paid for, and experts were to be highly qualified.

Along with the project, experts were provided with the questionnaire, developed by the consultant in cooperation with the Fund organizer (we will call him further a decision-maker, or DM). The questionnaire reflected the decision-maker's policy through the list of the most important for him criteria of the proposed R&D projects incorporated with the possible values on their scales. The list of these criteria with possible values was the following:

CRITERIA FOR R&D PROJECTS' EVALUATION

Criterion A. The existing basis for the project's goal

A1. There are unique manufactured examples
A2. There exists a production technology
A3. There exists only the idea

Criterion B. Period until expenses are paid

B1. Less than half a year after the beginning of the manufacturing of the product
B2. A year after the beginning of the manufacturing of the product
B3. Two or more years after beginning of the manufacturing of the product

Criterion C. Difficulties in the organization of the manufacturing process

C1. Not essential difficulties
C2. There are some difficulties
C3. There may be essential difficulties

Criterion D. Market situation

D1. There is a large demand for the product
D2. There is essential demand for the product
D3. The demand for the product is unknown

It is easy to note that criterion values are placed from the most preferred to the least preferred ones (according to the preferences of the decision maker).

The experts' evaluations of the projects (prepared in accordance with special rules) formed the basis for detailed analysis of the projects, and experts' criterion estimates were the basis for the projects' rank ordering. Let us note, that the expert was to assign one of the presented criterion values to each of the evaluated project. These estimates were used for computer analysis of the projects.

The consultant proposed to use method ZAPROS-LM for the task analysis. This method allows to select a group of better projects within the financial possibilities of the Fund.

It is not known in advance what projects (with what estimates combinations) will be submitted to the Fund. But in any case it is necessary to have means to rank-order the submitted projects according to their overall value. What is better in each case is a very subjective notion. Somebody has to measure the quality of the projects upon all criteria and make the necessary trade offs among them. As the decision maker is responsible for the Fund activity, his (or her) preferences are to be the basis for the projects' evaluation. It is necessary to reveal these preferences and construct the appropriate decision rule. Let us show how it is done in the method ZAPROS-LM.

Joint ordinal scale for two criteria

Let us look at the criteria' list. Assume that we have an "ideal" project, assigned all the best values against all criteria. We usually do not have such situation in real life. We will use this image as the reference point. Deviating from this ideal image, we will lessen the quality of the hypothetical project against two criteria: A ("The existing basis for the project's goal") and B ("Period until expenses are paid").

Question. What do you prefer: a project with the existing technology and a half year period until expenses are paid, or a project with the manufactured examples but with a year period until expenses are paid?

DM's answer. A project with the manufactured examples and a year period until expenses are paid is more preferable.

Question. What do you prefer: a project with the existing technology and a half year period until expenses are paid, or a project with the manufactured examples but with a two year or more period until expenses are paid?

DM's answer. A project with existing technology and a half year period until expenses are paid is more preferable

Question. What do you prefer: a project with the manufactured examples but with a two year or more period until expenses are paid, or project with a half year period until expenses are paid but with only the idea as the basis?

DM's answer. Both variants are poor in quality, but I would prefer the project with the existing manufactured examples.

The last question we will repeat once more under other assumptions about values against criteria C and D.

Figure 3.1. Comparison of criterion values for criteria A and B

A1B1		A1B1
↓		↓
A2B1	←	A1B2
↓	↘	↓
A3B1	←	A1B3

Question. Let assume that the demand for the project is unknown and there may be large difficulties in organizing the manufacturing of the product. What do you prefer: a project with the manufactured examples but with a two year or more period until expenses are paid, or project with a half year period until expenses are paid but with only the idea as the basis?

DM's answer. It is clear that both variants are very bad, but I would prefer the project with the existing manufactured examples.

In the figure 3.1 our questions and received answers are presented using criterion values denominations. Arrows mark the direction of preference.

The first and the second comparisons show that the values combination A2B1 may be placed between value combinations A1B2 and A1B3 according to the revealed decision maker's preferences. Then all the value combinations presented in the figure 3.1 may be rank-ordered and presented as a joint ordinal scale as follows:

A1B1C1D1 → A1B2C1D1 → A2B1C1D1 → A1B3C1D1 → A3B1C1D1

This joint scale may be presented in a more simple way if to take into account that for one of the two criteria A and B we have the best value, and that the unmarked here values against criteria C and D are also at the best level. In other words, instead of putting down A1B2C1D1 we will denote only B2 (the only value different from the best level). Then the constructed Joint Ordinal Scale may be presented as:

A1B1C1D1 → B2 → A2 → B3 → A3.

Thus, the answers to our questions allowed us to combine scales of criteria A and B into a joint scale. In the same way it is possible to combine scales of criteria A and C assuming the best values against criteria B and D, or to combine scales for criteria B and D assuming the best values against criteria A and C, and so on.

In other words, we work with all pairs of criteria, assuming the best values against all other criteria accept these two. The admissable operation 023 is used for the information elicitation (Ch. 2).

Let us formulate simple rules, how to formulate questions to combine scales of two criteria:

• two middle values are being compared; one of them becomes more preferable than the other;

• the less preferable one is compared with the worst value on the scale of the second criterion (in figure 3.1 it is seen that while comparing middle values B2 is preferred to A2; so, the next question is to compare A2 with B3);

• the less preferable criterion value in the second comparison is compared with the worst value of the second criterion (as B3 is compared with A3 in our example shown in figure 3.1).

Joint ordinal scale for all criteria

Let us assume that asking analogous questions and carrying out analogous comparisons we have constructed joint ordinal scales for all pairs of criteria (see figure 3.2).

Figure 3.2. Joint Ordinal Scales for all pairs of criteria

A1B1	A1C1	A1D1	B1C1	B1D1	C1D1
↓	↓	↓	↓	↓	↓
B2	A2	A2	B2	B2	C2
↓	↓	↓	↓	↓	↓
A2	C2	D2	C2	D2	D2
↓	↓	↓	↓	↓	↓
B3	A3	A3	B3	B3	C3
↓	↓	↓	↓	↓	↓
A3	C3	D3	C3	D3	D3

This gives us the easy possibility to construct the Joint Ordinal Scale (JOS) for all four criteria. As can be seen from the rank ordering of criterion values presented in figure 3.2, it is easy to determine their place in the mutual scale:

$$A1B1C1D1 \rightarrow B2 \rightarrow A2 \rightarrow C2 \rightarrow D2 \rightarrow B3 \rightarrow A3 \rightarrow C3 \rightarrow D3.$$

Verification of preference consistency

Comparisons carried out for one pair of criteria may be not consistent with comparisons carried out for the other pair of criteria. Suppose that the joint scale for criteria B and C is not the one presented in figure 3.2, but is as follows:

$$B1C1 \rightarrow C2 \rightarrow B2.$$

Then while trying to construct the joint scale for all criteria we will face contradictions. From the joint scale for criteria A and B it follows that B2 is preferred to A2, and from the joint scale for A and C it follows that A2 is preferred to C2 (see figure 3.2). Thus,

$$B2 \rightarrow A2 \rightarrow C2$$

But according to our assumption C2 is preferred to B2.

The current contradiction does not allow us to place A2, B2, and C2 on a joint scale. Usually this situation results from inconsistency in revealed preferences. It is necessary to analyze the carried out comparisons and make the necessary changes in them.

Thus, while constructing the joint scale for all criteria, there is a possibility to check for inconsistencies in the decision maker's responses. The possibility to combine several joint scales for pairs of criteria into a general joint scale confirms the consistency of decision maker's judgments.

Comparison of projects

Questions and answers, required for the construction of the joint ordinal scale, represent the only information needed from the decision maker for rank ordering of projects. In our case (4 criteria) the decision maker is to answer 24 questions (if there are no inconsistencies in the answers). Our experience in using ZAPROS computer system shows, that the dialogue for JOS construction takes about 10-15 minutes.

On the basis of the obtained information, the computer develops the decision rule for comparison of real projects. Let us explain how it is done.

First of all, all pairs of projects are being compared on the basis of the constructed Joint Ordinal Scale. Let assume there are two projects: N1 and N2.

Project N1 has criterion values: A2B2C1D1 (there exists a technology for manufacturing of the product, the period until expenses are paid is one year, the product is in great demand and there are no essential difficulties in organizing it's manufacturing).

Project N2 has criterion values: A1B2C2D2 (there exist manufactured examples of the product, the period until expenses are paid is one year, there is essential demand for the product, there are some difficulties in organizing it's manufacturing).

Using the Joint Ordinal Scale to compare these two projects we find out that: B2 is preferred to C2, and A2 is preferred to D2. Thus, project N1 is preferred to project N2 as it has more preferable combination of criterion values.

Let us note, that not always projects may be compared on the basis of the Joint Ordinal Scale. For example, project with criterion values A3B2C3D2 is incomparable with the project with criterion values A2B3C2D3, as B2 is preferred to A2 ad C3 is preferred to D3, but C2 is preferred to D2 and B3 is preferred to A3.

The system carries out paired comparisons for all pairs of projects, and then rank orders them in accordance with the obtained information. Let us show how it is done using an example.

Let have six projects: N1, N2, N3, N4, N5, N6. The results of paired comparisons are presented with an arrow from the more preferred to the less preferred object, no arrow means the objects are incomparable (see figure 3.3).

Figure 3.3. Paired comparisons of objects

$$N1 \nearrow N2 \searrow \quad N4 \to N5 \to N3$$
$$N1 \searrow N6 \nearrow$$

Let us select project N1 (not one arrow enters this object), and eliminate it from the set. Then projects N2 and N6 will be the best ones among the rest of objects (no arrows enter them). Eliminate them. Among the three objects left, N4 is the best one. And so on. Thus, we construct the rank ordering of alternatives:

$$N1 \to N2 \text{ and } N6 \to N4 \to N5 \to N6.$$

The advantages of the presented ZAPROS approach are:

1. all questions are simple and understandable for the decision maker. They are formulated using the language of criterion values;

2. it is possible to check decision maker's responses for consistency;

3. any comparisons of real projects may be explained using the same language of criterion values.

The analyzed task is one of the typical practical tasks facing the leaders of organizations dealing with competitive evaluation of proposals of different types and spheres. The proposed approach for these tasks' solution uses the preferences of the decision maker and is based on the method for rank ordering of multiattribute alternatives, named by the authors as ZAPROS-LM (Larichev & Moshkovich, 1995). It is necessary to note that this approach is a further development of ideas presented in (Gnedenko et al., 1986; Larichev et al., 1974; Larichev et al., 1978).

Problem formulation

There exists a rather large class of practical problems in which it is necessary to rank-order alternatives. The constructed alternatives' order may be used, e.g. to fund as many of the best projects as we can in such tasks as portfolio selection (Clarckson 1979, Furems & Moshkovich 1984), or to reject the least preferable alternatives from a production plan (Zuev et al., 1980; Larichev, 1982), or to define a limited group of the most preferred alternatives (Furems et al.,1982) and so on.

Let us assume that alternatives are estimated upon a set of criteria with verbal formulations of quality grades upon their scales (Larichev, 1979). An example of such criteria for the task of R&D evaluation are given in Table 3.1.

In cases when we have too many alternatives (at least dozens) it may be considered logical enough to construct some rules for pairwise comparison of alternatives on the basis of decision-maker (DM) preferences in the criteria space, and to use this set of rules for rank-ordering of the set of real alternatives. Just this problem is under further consideration.

The problem may be formulated as follows:

Given:
1. $K = \{q_i\}$ $i=1,2,...,Q$ is a set of criteria;

2. n_q is the number of possible values on the scale of the q-th criterion ($q \in K$);

3. $X_q = \{x_{iq}\}$ is a set of values for the q-th criterion (the scale of the q-th criterion); $|X_q| = n_q$ ($q \in K$);

4. $Y = X_1 * X_2 * ...* X_Q$ is a set of vectors $y_i \in Y$ of the following type

$$y_i = (y_{i1}, y_{i2},...,y_{iQ}); \text{ where } y_{iq} \in X_q \text{ and } N = |Y| = \prod_{q=1}^{Q} n_q;$$

5. $A = \{a_i\} \subseteq Y$ is a set of vectors, describing the real alternatives.

Required:

to form an ordering of multiattribute alternatives of the set A on the basis of the decision-maker's preferences.

An approach to the problem solution

As it has been mentioned above, we'll try to construct a rule for pairwise comparison of vectors from Y (on the basis of decision maker's preferences) and to apply this rule for comparison of vectors from A. Let us introduce a binary relation $R \subseteq Y \times Y$, reflecting relationship of preference or indifference in pairs of vectors from the set Y. Further the properties of the relation R which will be used, are marked (Litvak, 1982):

Table 3.1. Criteria and possible values for evaluation of R&D projects

Criteria	Possible values on their scales
1. Originality	1. Absolutely new idea and/or approach 2. There are new elements in the proposal 3. Further development of previous ideas
2. Prospects	1. High probability of success 2. Success is rather probable 3. There is some possibility of success 4. Success is hardly probable
3. Qualification	1. Qualification of the applicant is high 2. Qualification of the applicant is normal 3. Qualification of the applicant is unknown 4. Qualification of the applicant is low.
4. Level of the work	1. The proposed work is of high level 2. The proposed work is of middle level 3. The proposed work is of low level

<u>Definition 3.1.</u> R is called *reflexive*, if $(y_i, y_i) \in R, \forall y_i \in Y$.

Definition 3.2. R is called *connected*, if $\forall$ y_i, $y_j \in Y$ $(y_i,y_j) \in R$ or $(y_j,y_i) \in R$.

Definition 3.3. R is called *symmetric*, if $\forall$ y_i, $y_j \in Y$, $(y_i,y_j) \in R$ implies $(y_j,y_i) \in R$.

Definition 3.4. R is called *antisymmetric*, if $\forall$ y_i, $y_j \in Y$, $(y_i,y_j) \in R$ and $(y_j,y_i) \in R$ implies i=j.

Definition 3.5. R is called *transitive*, if $\forall$ y_i, y_j, $y_k \in Y$, $(y_i,y_j) \in R$ and $(y_j,y_k) \in R$ implies $(y_i,y_k) \in R$.

Definition 3.6. R is called a *quasi-order*, if R is reflexive and transitive.

Definition 3.7. R is called a *linear quasi-order*, if R is reflexive, transitive and connected.

Definition 3.8. P is called a *strict preference relation*, if it is antisymmetric and transitive.

Definition 3.9. I is called an *indifference* relation, if it is symmetric and transitive.

The most popular approach to solution of such task is the construction of a scalar value function v(y) with the following usual properties:

$v(y_i) > v(y_j)$, if $(y_i,y_j) \in P$;

$v(y_i) = v(y_j)$, if $(y_i,y_j) \in I$.

If a value function exists then it induces a complete ordering on the set Y. Conversely, if we do have a complete preference order on the set Y, then we are able to construct a value function by merely attaching any increasing sequence of numbers to vectors arranged in the increasing order.

The main idea of the approach described below is based on the concept of joint ordinal scale built according to the DM's preferences. The joint ordinal scale (JOS) means that all possible values upon all criteria are ranked-ordered for the DM upon his (or her) preferences. This ordinal scale may be effectively used for comparison of real alternatives.

Rather usual task, required from a decision maker in multicriteria decision problems with verbal criteria, is that of rank-ordering of possible values for one criterion from the set K (see, e.g., Goodwin & Wright, 1991; Keeney, 1992; von Winterfeldt & Edwards, 1986). As a result ordinal scales for criteria are formed, in which the first value x_{q1} upon the criterion q (q $\in$ K) is more preferable for a DM than the second value x_{q2} upon the same criterion and so on.

If we use natural numbers to enumerate values in the ordinal scale X_q for the q-th criterion, we shall obtain a modified ordinal scale $B_q=\{1,2,...,n_q\}$, where $b_{iq}<b_{jq}$, if x_{iq} is more preferable for the DM, than x_{jq}. So, for each ordinal scale X_q we form the unique ordinal scale B_q, reflecting the DM's preferences for values from X_q.

This information defines a relation of strict preference (or dominance) P^0 on the set Y:

$$P^0=\{ (y_i,y_j) \in Y \times Y | \forall q \in K\ b_{iq} \le b_{jq} \text{ and } \exists\ q^0 \text{ such that } b_{iq0}< b_{jq0}\}.$$

Let us now analyze what relation may be constructed on the set Y, if the joint ordinal scale according to the DM's preferences is formed. Let R be a linear quasi-order built on the set X = $\{X_q\}$ q=1,2,...,Q according to the DM's preferences. It is clear that the relations of strict preference P^0, defined on values of separate criteria, are the part of R (as they also reflect the DM's preferences). As in the previous case, we are able to enumerate elements from X in an increasing order according to R. Note that some elements from X will be assigned equal numbers as R is reflexive (that is there may be values equally preferable for the DM).

As now we know all the relations between values upon different criteria we are able to introduce the following binary relation of quasi-order on Y:

$$R^1=\{ (y_i,y_j) \in Y \times Y | \forall y_{iq} (q \in K) \exists\ y_{jt(q)}\ (t(q) \in K) \text{ such that } (y_{iq},y_{jt(q)}) \in R,$$

$$\text{and if } q \neq q_1, \text{ then } t(q) \neq t(q_1)\}.$$

It is clear that $P^0 \subset R^1$.

Thus the rule for comparison of vectors from Y (and from the set A accordingly) may be formulated as follows:

Definition 3.10. Vector $y_i \in Y$ is not less preferable than vector $y_j \in Y$, if for each component of the vector y_i there exists a component of the vector y_j with not more preferable value upon joint ordinal scale (binary relation R).

Thus, the task is to construct such joint ordinal scale on the basis of the DM's preferences. Further, it will be shown that to do this, we'll need simple ordinal pairwise comparisons, fulfilled by the DM for some vectors from Y, differing in values upon not more than two criteria. To prove the correctness of the introduced rule (definition 3.10), two rather simple assumptions about the properties of a decision-maker's preference system may be used: preference independence of criteria (Fishburn, 1970; Keeney & Raiffa, 1976) and transitivity of the resulting preference-indifference relation (Mirkin, 1974).

In the next two sections the way to elicit valid information on the DM's preferences is described. Verification of the transitivity of the DM's preferences is carried out, possible violations are corrected. In the following section foundations for the correctness of the implementation of JOS for comparison of vectors from Y are given. After that the possibilities for checking preference independence of criteria for the DM are discussed. Recommendations for task modification are given in the case of dependency.

Elicitation of information on DM's preferences

To construct the joint ordinal scale (according to it's definition) it is necessary to compare all possible pairs of values upon all criteria. To compare scale values of one criterion, it is enough to ask such questions as: which of these two values is more preferable for you? As has been stated before, this is an ordinary step, used in almost all decision making methods to construct an ordinal scale for each criterion. As a rule the decision maker does not have any problems responding to such questions, and the received judgments are rather stable.

Nevertheless, it is necessary to remember that it is usually assumed that the order of values for one criterion does not depend on the values upon other criteria (the axiom of preference independence is assumed). There may be examples, when this condition is violated. In such cases we recommend to use a combined criterion derived from the two dependent ones (see below).

Let assume that the ordinal criteria scales are formed and are reflected in scales B_q, described earlier. To carry out the comparison of values upon different criteria, it is necessary to ask the decision maker questions of the kind: "What do you prefer: to have the best level upon criterion q and the second (in the rank order) level upon criterion q+1, or the best level upon criterion q+1 and the second (in the rank order) level upon criterion q?".

The form of the question is rather difficult, besides, we have to bear in mind the possibility that values upon other criteria may influence the result of comparison. That's why we propose to compare vectors from Y which have all the same values but two. (For our case one alternative will have value 1 upon criterion q and value 2 upon criterion $q + 1$, and the second alternative will have value 2 upon criterion q and value 1 upon criterion $q + 1$).

Definition 3.11. The pair of vectors from Y will be called *admissible* for comparison if these vectors differ upon not more than two components.

In general the decision maker is to compare all admissible pairs of vectors from Y. But the number of such pairs from Y may be very large. Therefore it was proposed to compare vectors near two reference situations, as it will be shown that this information is enough to construct the joint ordinal scale.

Each vector from Y (that is a combination of values upon criteria) is an image of a certain alternative for a DM. The two most bright "contrasting" images correspond to the combinations of the best and the worst values upon all criteria. Such vectors were called reference situations (Larichev et al., 1978).

Definition 3.12. Vectors with all the best or all the worst values upon all criteria will be called *reference situations*.

Definition 3.13. Let us call *the list of vectors near a reference situation* a subset of vectors from Y with all components except one equal to those of this reference situation.

Let us form lists L_1 and L_2 near the first and the second reference situation correspondingly:

$L_1 = \{y_i \in Y \mid b_i = (1,1,1,...,1,b_{is},1,...1), b_{is} \neq 1, \forall s \in K \};$

$L_2 = \{y_i \in Y \mid b_i = (n_1,n_2,...,n_{s-1},b_{is},n_{s+1},...n_Q), b_{is} \neq n_s, \forall s \in K\}.$

It is clear that $|L_1| = |L_2| = N_1 = \sum_{q=1}^{Q}(n_q - 1)$.

For three criteria with three possible values on their scales, L_1 will consist of the following six vectors: (1,1,2), (1,1,3), (1,2,1), (1,3,1), (2,1,1), (3,1,1). Analogously the list L_2 will consist of : (3,3,1), (3,3,2), (3,1,3), (3,2,3), (1,3,3), (2,3,3).

We propose to carry out an interview with a DM for each list of vectors. The procedure will be described for the list L_1 as an example. An interview near the second reference situation (list L_2) is carried out in the same way.

The DM is asked to compare pairs of vectors from the list L_1 as it is proposed in (Gnedenko et al., 1986). All questions necessary to compare all vectors from the list L_1 are asked. The results of pairwise comparisons by a DM may be presented in a form of binary relations as follows:

1. $(y_i, y_j) \in P_{DM}$ if $(y_i, y_j) \in L_1 x L_1$, and according to a DM's opinion y_i is more preferable than y_j, or if $(y_i,y_j) \in P^0$.

2. $(y_i, y_j) \in P_{DM}^{-1}$ if $(y_i, y_j) \in L_1 x L_1$, and according to a DM's opinion y_j is more preferable than y_i, or if $(y_j,y_i) \in P^0$.

3. $(y_i, y_j) \in I_{DM}$ if $(y_i, y_j) \in L_1 x L_1$, and according to a DM's opinion y_i is equal to y_j, or if i=j.

As DM's answer: "vector y_i is more preferable than vector y_j", means that $(y_i,y_j) \in P_{DM}$ and $(y_j,y_i) \in P_{DM}^{-1}$, the relation P_{DM} is anti symmetric and anti reflexive.

As DM's answer: "vector y_i and vector y_j are equally preferable", means that $(y_i,y_j) \in I_{DM}$ and $(y_j,y_i) \in I_{DM}$, the relation I_{DM} is reflexive and symmetric.

If we require the transitivity of relations P_{DM} and I_{DM} then according to (Mirkin, 1974) the relation $R_{DM}=P_{DM} \cup I_{DM}$ is a linear quasi-order on the set L_1. Sequential elicitation of the necessary information allows to construct the Joint Ordinal Scale.

In any interview with a DM there is a possibility of errors in his (her) responses. These errors may be random or may occur while comparing similar in quality alternatives. Therefore, in the information, elicited from a DM inconsistencies (contradictions) may appear. A special procedure for detection and elimination of contradictions in DM's responses is proposed (Moshkovich, 1988).

Elimination of intransitivity in DM's responses

In the problem under consideration the possible contradictions in DM's responses may be determined as violations of transitivity of relations P_{DM} and I_{DM} (and in general as violations of transitivity of R_{DM}).

In general the problem of detection and elimination of intransitivity in pairwise comparisons is a rather complicated one. It is analogous to the task of cycles' elimination in a graph (Wilson, 1972; Kendall, 1969; Aho et al., 1962; Ore, 1962). It is known that the problem of determination of the minimal number of arcs necessary to be destroyed in a graph to make it acyclic is a NP-complete problem (Garey & Johnson, 1979). This means that in nowdays it is considered that this problem can not be solved exactly in a polynomial time. That is why there are works, devoted to the development of an approximate solution of this problem (Aho et al., 1974; Ore, 1962).

Our problem of detection and elimination of intransitivity of the information, received from a DM, has two peculiarities which make traditional approach above mentioned ineffective.

First, the elimination of arcs in a graph may lead to a partial loss of information on vectors' comparisons which is undesirable. Secondly, our task is to detect the erroneous DM's responses which have led to cycles, but not to find the minimal number of arcs to be eliminated.

The main idea of the proposed approach (Moshkovich, 1988) is as follows. It is based on the assumption of the transitivity of DM's preferences and considers violations of this assumption to be errors in DM's responses.

The transitivity of preferences assumes that if:

1. $(y_i, y_j) \in P_{DM}$, then $\forall y_k \in L_1$ and $(y_j, y_k) \in P_{DM}$, $(y_i, y_k) \in P_{DM}$;

2. $(y_i, y_j) \in I_{DM}$, then $\forall y_k \in L_1$ and $(y_j, y_k) \in I_{DM}$, $(y_i, y_k) \in I_{DM}$;

3. $(y_i, y_j) \in P_{DM}$, then $\forall y_k \in L_1$ and $(y_j, y_k) \in I_{DM}$, $(y_i, y_k) \in P_{DM}$;

4. $(y_i, y_j) \in I_{DM}$, then $\forall y_k \in L_1$ and $(y_j, y_k) \in P_{DM}$, $(y_i, y_k) \in P_{DM}$.

Therefore after each comparison of vectors from L_1 made by a DM, this information may be extended on the basis of transitivity (transitive closure of the binary relation defined on the set L_1 is being built).

After that the DM is presented with the next pair of vectors from L_1, for which the relation has not been defined. When the DM's response is obtained, the transitive closure is developed and the procedure is maintained up to the moment of establishing relations for all pairs from L_1.

Statement 3.1. If $R_{DM} = P_{DM} \cup I_{DM}$ is transitive and $(y_i, y_j) \notin R_{DM}$, then the *transitive closure* R^*_{DM} of the relation $R_{DM} := R_{DM} \cup (y_i, y_j)$ will be transitive for any type of the DM's response on comparison of y_i and y_j

The proof is evident because a DM is presented only with pairs of vectors from $L_1 \times L_1$, for which previous responses have not predefined any relation. So, any variant of the response $((y_i, y_j) \in P_{DM}; (y_i, y_j) \in I_{DM}; (y_j, y_i) \in P_{DM})$ will not contradict previous responses).

Once the response is received, transitive closure of the newly obtained relation is being built. It is known that transitive closure of the acyclic graph does not lead to cycles (Aho et al., 1974). So, we can say that such a procedure does not lead to intransitivity of the relation being built.

On the other hand, using the transitive closure to add information on comparisons assumes that the DM's responses do not contain errors. As it is necessary to verify judgments obtained from people, it is proposed to present the

DM with additional pairs of vectors for comparison on the basis of the following principle:

the relation between each pair of vectors from L_1 is to be defined directly (by a DM's response) or indirectly (by transitive closure) no less that two times.

This requirement means that if a DM by two of his (her) responses (may be indirectly - by transitive closure) has equally defined the relation between vectors from L_1 in some pair, then this relation is considered to be proven. If the relation between vectors from L_1 in some pair has been defined only once and only upon transitive closure, then this pair is presented additionally to a DM for comparison.

If the DM's response does not conflict with the previously obtained information, then the judgment is considered to be correct. If there is some difference, the triple of vectors for which a pairwise comparison contradicts the transitivity of the relation being built on L_1, is found out: that is of vectors y_i, y_j, $y_k \in L_1$ such that one of the following statements is fulfilled:

1. $(y_i,y_j) \in P_{DM}$; $(y_j,y_k) \in I_{DM}$; $(y_i,y_k) \in I_{DM}$;

2. $(y_i,y_j) \in P_{DM}$; $(y_j,y_k) \in P_{DM}$; $(y_i,y_k) \in I_{DM}$;

3. $(y_i,y_j) \in P_{DM}$; $(y_j,y_k) \in P_{DM}$; $(y_k,y_i) \in P_{DM}$.

Such triple may always be detected, because after each of DM's responses we have built transitive closure of the obtained relation. In this case the DM is asked to reconsider the situation and to change one (or more) of his (or her) previous responses to eliminate intransitivity.

After the corrected responses are obtained, they are incorporated into the information on the DM's preferences as follows.

It is supposed that we only start the interview with a DM (that is we have only these three responses for pairwise comparisons of y_i with y_j; y_j with y_k; y_k with y_i). At this time we also know that these responses do not contradict each other. We assign each of these responses to P_{DM} or I_{DM} accordingly and carry out the transitive closure of the obtained relation as in the initial interview with a DM. Subsequently we carry out further formation of the binary relation on L_1.

Information for pairwise comparisons is obtained from the previous responses of a DM. This is again followed by transitive closure of the relation R_{DM}.

This way guarantees that previous DM's responses do not contradict the newly built relation (as we use only responses for those pairs of vectors for which previous responses have not predefined some relation). As a result we obtain new transitive relation on the set L_1 in which the necessary changes have been made, but all previous responses not contradictory to the new ones are maintained unchanged. After that, the condition of "double test" for each pairwise comparison is checked for this new information.

The proposed approach makes it possible to form an effective procedure for an interview with a DM to build the required relation, as the redundancy of the obtained information is limited to a reasonable condition of minimal necessary test for DM's responses.

Implementation of information on the DM's preferences

As a result of an interview with a DM and transforming his (her) responses to a non-contradictory variant the relation $R_1 = P_{DM} \cup I_{DM}$ of a linear quasi-order on the set L_1 is built.

Analogously, the binary relation R_2 on the set L_2 (near the second reference situation) is being built.

As a result we obtain a certain amount of information on the decision maker's preference system and the question of it's effective use arises.

Preference independence of criteria and its implementation

To use effectively the information obtained from a DM it is necessary to have preference independence of all pairs of criteria (Keeney, 1974).

Definition 3.13. Criteria s and t of the set K are *preferentially independent from the other criteria of this set*, if preference between vectors with equal values upon all criteria but s and t, does not depend on the values of the equal components.

In practical problems we must check if this axiom is not violated in DM's preferences. The problem of checking this axiom (as well as checking many other axioms of multiattribute utility theory) has no simple solution. In reality, the necessity to use this axiom results from the desire to construct an effective decision rule on the basis of relatively small amount of rather simple information about DM's preferences (the effectiveness of the decision rule means its possibility to guarantee rather high level of comparability for real alternatives). On the other hand, the full-scale check of DM's preferences implies the need for a DM to carry out a large number of pairwise comparisons. So, the point is to make not a full-scale but sufficient check of DM's preferences to satisfy the axiom's conditions. There exist several approaches to these problem (Gnedenko et al., 1986; Keeney, 1980; Larichev et al., 1979;von Winterfeldt & Edwards, 1986).

Approaches to axiom's verification

Keeney (1980) suggested to carry out the axiom's checking on a set of specially constructed pairs of alternatives. Two criteria s and t were selected. Components for all other criteria for both alternatives were fixed at the least preferable level. Then the decision maker was asked to mark the indifference levels for criteria s and t $((y_{is}, y_{jt})$ and $(y_{is}^{*}, y_{jt}^{*}))$ in a way for the following two vectors to be equally preferable:

$$(n_1, n_2, ..., n_{s-1}, b_{is}, n_{s+1}, \quad , n_{t-1}, b_{jt}, n_{t+1}, ..., n_Q)$$

$$(n_1, n_2, ..., n_{s-1}, b_{is}^{*} \ n_{s+1}, \quad , n_{t-1}, b_{jt}^{*}, n_{t+1}, ..., n_Q).$$

Then the values for fixed components in these two vectors were being changed onto the most preferable ones and the decision maker was to state if these vectors were equally preferable for him (or her). If for several different pairs of values against criteria s and t that condition was fulfilled, the conclusion was made that criteria s and t were preferentially independent from the other criteria.

Dialogue with the decision maker described in the work of Keeney (1980) shows the difficulty of such work for the decision maker. von Winterfeldt

(1975) notes that the process of seeking indifference points increase errors in decision maker's responses.

Gnedenko with colleagues (1986) proposed to carry out analogous procedure on the basis of comparison of pairs of vectors from a special list L_3:

$$L_3 = \{y_i \in Y | b_i = (n_1, n_2, \ldots, n_{s-1}, b_{is}, n_{s+1}, \ldots, n_{t-1}, b_{it}, n_{t+1}, \ldots, n_Q), b_{is} \neq 1, b_{it} = 1 \text{ for} \forall s, t \in K, s \neq t\}$$

Then for each pair (y_i, y_j) from list L_1 it is possible to find pair $(y_i^*, y_j^*) \in L_3$ in the following way.

Let y_i be different from y_j against criteria s and t, and (according to the peculiarity of L_1) only one of these two components will be different from 1. Let $b_{is} = 1$, and $b_{jt} = 1$. This means:

$$b_i = (1, 1, \ldots, 1, b_{it}, 1, \ldots, 1); \quad b_j = (1, 1, \ldots, 1, b_{js}, 1, \ldots, 1).$$

Then $y_i^*, y_j^* \in L_3$ will be as follows:

$$b_i^* = (n_1, n_2, \ldots, n_{s-1}, 1, n_{s+1}, \ldots, n_{t-1}, b_{it}, n_{t+1}, \ldots, n_Q),$$

$$b_j^* = (n_1, n_2, \ldots, n_{s-1}, b_{js}, n_{s+1}, \ldots, n_{t-1}, 1, n_{t+1}, \ldots, n_Q)$$

It is evident that if criteria s and t are preferentially independent from the other criteria from K, the relation between vectors (y_i, y_j) and (y_i^*, y_j^*) is to be the same, as these pairs of vectors differ only in the level of fixed components.

Thus, if these relations are the same, this fact supports the preference independence of criteria s and t. If the relations are different, then, it's a good chance that they are not independent.

Of course, this verification is not complete, but if we consider the fact that fixed components upon Q-2 criteria are changed from the least preferable to the most preferable ones, it is possible to consider that if in these circumstances the decision maker's preferences are the same for all such pairs of vectors, the criteria s and t are preferentially independent from the others.

What is the problem with such an approach?

First, the list L_3 is of the size of lists L_1 and L_2, together, though it plays only subordinate role. If the number of elements in L_1 (equal to the number of elements in L_2) is equal to $N_1 = \sum_{q=1}^{Q}(n_q - 1)$, the number of elements in L_3 (as it was shown in (Larichev & Moshkovich, 1995), is equal to

$$|L_3| = N_3 = (Q-1)\sum_{q=1}^{Q}(n_q - 1) - Q(Q-2).$$

This is seen from the fact that for each criterion q we form $n_q - 1$ vectors with only q-th component being different from n_q and with the most preferable value for component t. Then for t=1,2,...,Q and t≠q, the number of such vectors would be equal to $(Q - 1)(n_q - 1)$.

As a result for all q=1,2,...,Q the number of corresponding vectors will be equal to $(Q - 1)\sum_{q=1}^{Q}(n_q - 1)$.

In this estimate vectors which have one most preferable component and others are the least preferable ones are calculated several times. To make the estimation exact we have to use the previously introduced formula for list L_3.

Second, type of vectors in L_3 differ considerably from the ones from lists L_1 and L_2. Vectors from L_1 and L_2 are different from the reference situation in only one component, in L_3 we have vectors differing from the reference situation by two components. Thus, their comparison requires different cognitive efforts from the decision maker.

Third, it is not possible to present the decision maker with all components from L_3 (as they may differ in three components). Thus, it is not possible to construct a connected binary relation on the set L_3 and to check decision maker's responses on the basis of transitivity. In this case the possible erroneous decision maker's response may lead to incorrect conclusion about the independency of criteria under consideration.

All these led to the necessity to modify the process of axiom verification. The following approach was proposed.

Let us recall that part of vectors from list L_3 are also included into list L_2 (those that have all component but one on the least preferable level). There is a possibility to compare relations between pairs of vectors near two reference situations of the following type:

L_1: y_i such that $b_i = (1, 1,..., 1, n_s, 1,..., 1)$

  y_j such that $b_j = (1, 1,..., 1, n_t, 1,..., 1)$

L_2: y_i^* such that $b_i^* = (n_1, n_2, ..., n_{s-1}, n_s, n_{s+1}, ..., n_{t-1}, 1, n_{t+1}, ..., n_Q)$

  y_j^* such that $b_j^* = (n_1, n_2, ..., n_{s-1}, 1, n_{s+1}, ..., n_{t-1}, n_t, n_{t+1}, ..., n_Q)$

Both pairs of vectors differ only in components upon criteria s and t. So, pairs differ from one anther only in values of equal components. Therefore, if criteria s and t are preferentially independent, the preference in the pairs (y_i, y_j) and (y_i^*, y_j^*) has to be the same. Thus, there is a possibility to carry out some justification of the axiom on the basis of the information obtained near two reference situations.

Let us emphasize that though such justification is a very limited one, the violation of this condition rather clearly proves the violation of independence and the necessity of additional analysis of the situation (see later), as all these relations have been thoroughly checked during comparisons near two reference situations. Additionally, let us note that the selected reference situations differ to a very large extent, so the correspondence of the results obtained near them, may be considered to be stable and for all intermediate situations.

Implementation of obtained information for comparison of real alternatives

Information on comparison of vectors' pairs obtained near the first reference situation may be used for construction of the joint ordinal scale.

The joint ordinal scale in (Larichev et al.,1974; Ozernoy & Gaft, 1978) was considered to be the ranking of the set of vectors near the first reference situation. If the binary relation built on the set L_1 is connected, then we have a complete (linear) quasi-order on this set. Thus, each vector in this quasi-order may be assigned the number of it's place in this ranking.

If we recall that vectors near the first reference situation differ from that reference situation in only one component, we can consider the place, obtained by the vector in this ranking to be the place of this unique component in the Joint Ordinal Scale (JOS).

In (Gnedenko et al., 1986) a rule for comparison of any vectors on the basis of the joint ordinal scale was formulated. The correctness of the rule may be proved for the case of pairwise preference independence of all criteria (Gnedenko et al. 1986, Larichev & Moshkovich, 1991).

Now, maintaining the formal adequacy of the statement, given in (Gnedenko et al. 1986), let us formulate it in a more precise way.

Let $L'=L_1 \cup (1, 1,..., 1)$ (L_1 is complemented by a vector with all the best values). The relation R_1 is complemented by relations which reflect the preference of the vector $(1, 1,..., 1)$ to all other vectors from L_1 and its equality to itself. Then the statement may be formulated in the following way.

Statement 3.2. *If each pair of criteria from K (Q>3) does not depend preferentially on other criteria, then vector $y_i=(y_{i1},y_{i2},...,y_{iQ}) \in Y$ is not less preferable for the DM than vector $y_j=(y_{j1},y_{j2},...,y_{jQ}) \in Y$, if for each criterion $s \in K$ there exists criterion $t(s) \in K$ such that:*
$(1, 1, ..., 1, b_{is}, 1, ... ,1) R_1 (1, 1, ..., 1,b_{jt(s)}, 1,..., 1)$
and if $s \neq q$ then $t(s) \neq t(q)$.

Proof of the statement. According to the corollary from theorem 3.7 (Keeney & Raiffa, 1976), if each pair of criteria is independent of it's supplement, the criteria are mutually preferentially independent. Thus, according to theorem 3.6 (Keeney & Raiffa, 1976) for criteria from K there exists an additive value function $v(y_i) = \sum_{q=1}^{Q} v_q(y_{iq})$.

Let there be two vectors from Y: $y_i = (y_{i1}, y_{i2}, ..., y_{iQ})$ and $y_j = (y_{j1}, y_{j2}, ..., y_{jQ})$.

Then according to the statement for each y_{is} there exists y_{jt} such that when all other components in vectors are the most preferred ones, then y_{is} is not less preferable than y_{jt}. Let have it as follows:

1) $(b_{i1}, 1, 1, ..., 1)\ R_1\ (1, b_{j2}, 1, ..., 1)$;
2) $(1, b_{i2}, 1, ..., 1)\ R_1\ (1, 1, b_{j3}, 1, ..., 1)$;
   ....
Q) $(1, 1, ..., 1, b_{iQ})\ R_1\ (b_{j1}, 1, 1, ..., 1)$.

Expression 1) means that

$$v_1(b_{i1}) + v_2(1) + ... + v_Q(1) \geq v_1(1) + v_2(b_{j2}) + ... + v_Q(1).$$

Thus, $v_1(b_{i1}) + v_2(1) \geq v_1(1) + v_2(b_{j2})$.

Expression 2) means that

$$v_1(1) + v_2(b_{i2}) + ... + v_Q(1) \geq v_1(1) + v_2(1) + v_3(b_{j3}).... + v_Q(1).$$

Thus, $v_2(b_{i2}) + v_3(1) \geq v_2(1) + v_3(b_{j3})$ and so on.

Adding up left and rights parts of inequalities, we get:

$$v_1(b_{i1}) + v_2(1) + v_2(b_{i2}) + v_3(1)... + v_{Q-1}(b_{iQ-1}) + v_Q(1) + v_Q(b_{iQ}) + v_1(1) \geq v_1(1) +$$
$$+ v_2(1) + v_2(b_{j2}) + .v_3(b_{j3}) + ... + v_{Q-1}(1) + v_Q(b_{jQ}) + v_Q(1) + v_1(b_{j1}).$$

Eliminating equivalent elements in both parts we get:

$$v_1(b_{i1}) + v_2(b_{i2}) + ... + v_{Q-1}(b_{iQ-1}) + v_Q(b_{iQ}) \geq v_2(b_{j2}) + .v_3(b_{j3}) + ... + v_Q(b_{jQ}) + v_1(b_{j1}).$$

Thus, $v(y_i) \geq v(y_j)$ and this is the required result.

It is easy to prove that the introduced rule for comparison of vectors from Y, may be modified for implementation on the basis of the relation R_2 constructed near second reference situation.

Statement 3.3. *If each pair of criteria from K (Q>3) does not depend preferentially on other criteria, then vector $y_i=(y_{i1},y_{i2},...,y_{iQ}) \in Y$ is not less preferable for the DM than vector $y_j=(y_{j1},y_{j2},...,y_{jQ}) \in Y$, if for each criterion s ∈ K there exists criterion t(s) ∈ K such that:*

$(n_1, n_2, ..., n_{s-1}, b_{is}, n_{s+1}, ... , n_q) R_2 (n_1, n_2, ..., n_{t(s)-1}, b_{jt(s)}, n_{t(s)+1}, ... , n_q)$
and if s≠q then t(s) ≠ t(q).

The proof of the statement is analogous to the previous one.

It is somewhat more complicated to prove that when criteria are pairwise independent, then y_i is not less preferable than y_j, if for part of components we can find not less preferable ones from L_1, and for the other part - from L_2. In this case additional requirement for the subsets of criteria for elements from relations R_1 and R_2 is to be met.

Statement 3.4. *If each pair of criteria from K (Q>3) does not depend preferentially on other criteria, then vector $y_i=(y_{i1},y_{i2},...,y_{iQ}) \in Y$ is not less preferable for the DM than vector $y_j=(y_{j1},y_{j2},...,y_{jQ}) \in Y$, if:*

1) for each criterion s ∈ $K_1 \subset K$ there exists criterion t(s) ∈ K_1 such that:
$(1, 1, ..., 1, b_{is}, 1, ... ,1) R_1 (1, 1, ..., 1, b_{jt(s)}, 1,..., 1)$
and if s≠q then t(s) ≠t(q).

2) for each criterion s ∈ $K_2 = K \backslash K_1$ there exists criterion t(s) ∈ K_2 such that: $(n_1, n_2, ..., n_{s-1}, b_{is}, n_{s+1}, ... ,n_Q) R_2 (n_1, n_2, ..., n_{t(s)-1}, b_{jt(s)}, n_{t(s)+1}, ... ,n_Q)$
and if s≠q then t(s) ≠t(q).

Proof of the statement 3.4. As in the previous case, pairwise criteria independence defines the existence of the additive value function

$$v(y_i) = \sum_{q=1}^{Q} v_q(y_{iq}).$$

Let there be two vectors from Y: $y_i=(y_{i1},y_{i2},...,y_{iQ})$ and $y_j=(y_{j1},y_{j2},...,y_{jQ})$.

Then, according to the statement for each y_{is} there exist y_{jt} such that 1) or 2) from the statement is fulfilled.

Let consider that 1) is fulfilled for the first m criteria, and 2) is fulfilled for all other criteria. This means that:

1) $(b_{i1}, 1, 1,...,1) R_1 (1,1,...1, b_{jt(1)}, 1,....,1)$;
2) $(1, b_{i2}, 1,...,1) R_1 (1, 1,...1, b_{jt(2)}, 1,....,1)$;

....

m) $(1,1,...,1, b_{im}, 1,...,1) R_1 (1, 1,...1, b_{jt(m)}, 1,....,1)$;
m+1) $(n_1, n_2, ..., n_m, b_{i(m+1)}, n_{m+2}, ... ,n_Q) R_2 (n_1, n_2, ..., b_{jt(m+1)}, n_{t(m+1)}, ... ,n_Q)$

....

Q) $(n_1, n_2, ...,n_{Q-1}, b_Q) R_2 (n_1, n_2, ..., b_{jt(Q)}, n_{t(Q)}, ... ,n_Q)$

As in the previous case, these expressions may be presented as inequalities of sums of value functions. Summing them up we get:

$$\sum_{s=1}^{m}\sum_{q\neq s}^{m} v_q(1) + \sum_{q=1}^{m} v_q(b_{iq}) + \sum_{s=m+1}^{Q}\sum_{q\neq s}^{Q} v_q(n_q) + \sum_{q=m+1}^{Q} v_q(b_{iq}) \geq \sum_{s=1}^{m}\sum_{q\neq t(s)}^{m} v_q(1) +$$

$$+ \sum_{s=1}^{m} v_{t(s)}(b_{jt(s)}) + \sum_{s=m+1}^{Q}\sum_{q\neq t(s)}^{Q} v_q(n_q) + \sum_{s=m+1}^{Q} v_{t(s)}(b_{jt(s)}).$$

This leads to

$$\sum_{s=1}^{m}\left(\sum_{q\neq s} v_q(1) - \sum_{q\neq t(s)} v_q(1)\right) + \sum_{q=1}^{Q} v_q(b_{iq}) + \sum_{s=m+1}^{Q}\left(\sum_{q\neq s} v_q(n_q) - \sum_{q\neq t(s)} v_q(n_q)\right) \geq \sum_{q=1}^{Q} v_q(b_{jq})$$

As a result we obtain:

$$\sum_{s=1}^{m}\left(v_{t(s)}(1) - v_s(1)\right) + \sum_{q=1}^{Q} v_q(b_{iq}) + \sum_{s=m+1}^{Q}\left(v_{t(s)}(n_{t(s)}) - v_s(n_s)\right) \geq \sum_{q=1}^{Q} v_q(b_{jq}).$$

Note, that the first and the last sums in the left part of inequality are equal to zero as according to the statement $t(s) \in \{1,2,...,m\}$ when $s=1,2,...,m$; and

$t(s) \in \{m+1,m+2,...,Q\}$ when $s=m+1,m+2,...,Q$. At the same time when $q \neq s$ then $t(q) \neq t(s)$.

Thus, $v(y_i) \geq v(y_j)$ (as it was required).

What do we get out of these results?

While comparing real alternatives it is possible to use joint ordinal scales, built near both reference situations. If the results are the same, then this is an indirect confirmation of preference independence of criteria and justification of rules being used (this means that we carry out additional check of criteria independence just while comparing real alternatives).

If according to one rule alternatives can be compared and according to the other we are not able to compare them, then this is not a contradictory situation. We have just enlarged compatibility of alternatives on the basis of additional information from a DM about comparison of vectors from L_1.

If the results of comparison contradict each other, then this is connected with violation of criteria independence. The alternatives are to be considered incomparable.

Even if there is some evidence about the dependency of some pair of criteria, we are able to use the built rules for comparison the pairs of alternatives, uninfluenced by this dependency. We are able to estimate the number of alternatives' pairs which it will be possible to compare additionally if we analyze the dependency thoroughly. Analysis and elicitation of dependent criteria and also procedures for reformulation of the initial task in this case (Larichev & Moshkovich, 1991) are rather labor-consuming.

In this case the decision maker is able to evaluate if he (or she) wants to spend rather large amount of time and effort, knowing the maximum of additional information about comparison of alternatives which it is possible to obtain as a result.

Task modification in case of criteria dependency

It is clear that differences in comparison of vectors from Y on the basis of the JOS built near two reference situations, are caused only by the information about DM's preferences presented in relations R_1 and R_2. Let us prove the following statement.

Statement 3.5. *Let the comparison of vectors $(y_i, y_j) \in Y$ on the basis of the relation R_1 and the relation R_2 be different. Then there always exist criteria s and t for which:*

$(1, 1, ..., 1, b_{is}, 1, ... , 1) R_1 (1, 1, ..., 1, b_{jt(s)}, 1, ..., 1)$ but
$(n_1, n_2, ..., n_{t(s)-1}, b_{jt(s)}, n_{t(s)+1}, ... , n_Q) R_2 (n_1, n_2, ..., n_{s-1}, b_{is}, n_{s+1}, ... , n_Q)$
(if $s \neq q$ then $t(s) \neq t(q)$).

Proof of the statement 3.5. As $(y_i, y_j) \in R_1$, then for each $s \in K$ there exists $t(s) \in K$ such that $(1, 1, ..., 1, b_{is}, 1, ... , 1) R_1 (1, 1, ..., 1, b_{jt(s)}, 1, ..., 1)$. As R_2 is a connected binary relation we have either

(1) $(n_1, n_2, ..., n_{s-1}, b_{is}, n_{s+1}, ... , n_Q) R_2 (n_1, n_2, ..., n_{t(s)-1}, b_{jt(s)}, n_{t(s)+1}, ... , n_Q)$

or

(2) $(n_1, n_2, ..., n_{t(s)-1}, b_{jt(s)}, n_{t(s)+1}, ... , n_Q) R_2 (n_1, n_2, ..., n_{s-1}, b_{is}, n_{s+1}, ... , n_Q)$.

If (2) is true that the statement is proven. Let assume (1) is true for every $s \in K$. Then according to the rule $(y_i, y_j) \in R_2$. But this contradicts to the statement.

This statement allows us to carry out the check of the preference independence of criteria on the basis of R_1 and R_2 due to the following corollaries.

Corollary 3.1. *If $y_i, y_j \in Y$ and comparison of y_i and y_j are different upon R_1 and R_2, then there always may be found criteria s and t such that:*
$(1, 1, ..., 1, b_{is}, 1, ... , 1) R_1 (1, 1, ..., 1, b_{jt}, 1, ..., 1)$ and either
$(n_1, n_2, ..., n_{t-1}, b_{jt}, n_{t+1}, ... , n_Q) R_2 (n_1, n_2, ..., n_{s-1}, b_{is}, n_{s+1}, ... , n_Q)$, or
$((n_1, n_2, ..., n_{t-1}, b_{jt}, n_{t+1}, ... , n_Q), (n_1, n_2, ..., n_{s-1}, b_{is}, n_{s+1}, ... , n_Q)) \notin R_2$.

Proof is evident.

Let us introduce the notion of analogous pairs.

Definition 3.15. Let us call a pair of vectors $(y_i', y_j') \in L_2$ analogous to the pair of vectors $(y_i, y_j) \in L_1$, if:

$y_i = (1, 1, ..., 1, b_{is}, 1, ... , 1);$ $y_j = (1, 1, ..., 1, b_{jt}, 1, ..., 1)$

$y_i' = (n_1, n_2, ..., n_{s-1}, b_{is}, n_{s+1}, ... , n_Q)$ $y_j' = (n_1, n_2, ..., n_{t-1}, b_{jt}, n_{t+1}, ... , n_Q).$

Corollary 4.2. If comparisons between all analogous pairs of vectors near two reference situations are the same, then it is impossible to detect violations of preference independence of criteria on the basis of the obtained information.

The same comparison of analogous pairs of vectors means that:

$$\text{if } (1, 1, ..., 1, b_{is}, 1, ... , 1)\ R_1\ (1, 1, ..., 1, b_{jt}, 1, ..., 1)$$
$$\text{then}$$
$$(n_1, n_2, ..., n_{s-1}, b_{is}, n_{s+1}, ... , n_Q)\ R_2,\ (n_1, n_2, ..., n_{t-1}, b_{jt}, n_{t+1}, ... , n_Q).$$

(The proof is evident, if to consider the above marked possibilities to detect violations of the axiom about preference independence of criteria).

This gives us the opportunity to find out pairs of dependent criteria by analyzing comparisons near two reference situations in the following way.

1.   The pair of dependent criteria are elaborated.

Criteria s and t are defined for which the comparison of analogous pairs of vectors near two reference situations is different. After that the decision maker is to mark what criterion value has changed the result. If the decision maker refuses to mark such criterion, one criterion (different from s and t) is selected at random. The value upon this criterion in the analogous pair of vectors near the second reference situation is changed for the most preferable one, and the decision maker is to compare these modified vectors. If the result of the comparison is the same as near the first reference situation, it is logical to make the conclusion that this criterion value has caused the difference in the previous case. If the result is different the same actions are used for the next criterion and so on.

As a result the criterion q and value n_q influencing the change in preferences near the second reference situation are defined.

2. The reason for the elaborated dependency is analyzed.

The difference in preferences near two reference situations may be caused by one of the two reasons (violations of initial assumptions).

First, it is possible that the rank order of values upon criterion s and/or t is being changed if values upon criterion q are changed. That's why it is necessary to check if the rank order of values upon criterion s and t are maintained for different values upon criterion q. To check this the decision maker is presented with pairs of vectors of the type:

$$b_i = (1, 1, ..., 1, b_{is}, 1, ... ,1,n_q,1,...,1) \text{ and } b_j = (1, 1, ..., 1, b_{js}, 1, ... ,1,n_q,1,...,1).$$

If at least for one criterion (e.g., s) the rank order of values changes with changing value upon criterion q it means criteria s and q are mutually dependent. It is proposed to combine these two criteria into one with scale containing in general case, all possible combinations of values upon criteria s and q.

Second, if the rank order is stable for both criteria s and t, then the conclusion is made that all three criteria (s, t, and q) are mutually dependent. In this case the combined scale for all three criteria is to be constructed (see Ch.2).

3. The scale for the new criterion is being built.

The decision maker is proposed to rank order the combinations of values upon dependent criteria. In general there may be equally preferable combinations. These cases may allow to decrease the number of values on the scale of a new criterion by reformulating equally preferable combinations into one value.

4. New preference elicitation process is carried out (as in previous case) as the set of criteria has been changed.

In this way it is possible to detect groups of dependent criteria, construct new combined scales for them, and to rank order real alternatives according to the proposed method.

Note, that the complexity of the task will be reduced as the number of criteria will be decreased.

Let us analyze the proposed method using a simple example.

Illustration of the method ZAPROS-LM using a simple example

Let us illustrate the proposed method ZAPROS-LM on a simple example. A young man, about to graduate from the college, receives concrete job offers from three companies, whose sites he has visited. He plans to accept one of the three jobs, but is not sure which he prefers most. He uses three main criteria to characterize each proposal: SALARY, JOB LOCATION, TYPE OF JOB POSITION. He estimated the proposals upon these criteria as shown in table 3.2.

Table 3.2: Description of alternatives using criteria

Jobs	Criteria		
	SALARY	JOB LOCATION	JOB POSITION
Proposal 1	Rather high	Some distance	Inappropriate
Proposal 2	Average	Far away	Ideal
Proposal 3	Lower	Ideal	Good enough

We can say that we need to compare vectors $b_1=(1,2,3)$; $b_2=(2,3,1)$ and $b_3=(3,1,2)$ which correspond to proposal 1, proposal 2 and proposal 3 respectively. Let us try to do it through construction of the joint ordinal scale for the considered three criteria.

First we form list of vectors near the first reference situation $L_1=\{211,311,121,131,112,113\}$. Thus, we have six vectors in this list and need to fill in the matrix of pairwise comparisons for them (further we'll show only the elements above the diagonal). We will use the following denominations: 1 to mark that the element in the row is more preferable than element in the column; 2 - to show equally preferable elements; 3 - to mark that the element in the row is less preferable than the element in the column; 0 - means that these elements are not compared.

Let present the initial matrix (A) with comparisons made upon the relation of dominance P^0.

Matrix A   Matrix B

	211	311	121	131	112	113	211	311	121	131	112	113
211	2	1	0	0	0	0	2	1	1	$\underline{1}$	0	0
311		2	0	0	0	0		2	0	0	0	0
121			2	1	0	0			2	1	0	0
131				2	0	0				2	0	0
112					2	1					2	1
113						2						2

In (B) you can see the same matrix, but after the first comparison fulfilled by he DM: he prefers vector 211 to vector 121. The underlined 1 marks the conclusion made upon transitivity: $(211,121) \in P_{DM}$ and $(121,131) \in P_{DM}$. This implies that $(211,131) \in P_{DM}$.

Now the DM is to compare vectors 211 and 112. Matrix (C) represents the entirely fulfilled matrix (underlined numbers mean that they are defined upon transitivity).

Matrix C

	211	311	121	131	112	113
211	2	1	1	$\underline{1}$	1	$\underline{1}$
311		2	$\underline{3}$	3	3	1
121			2	1	3	$\underline{1}$
131				2	$\underline{3}$	1
112					2	1
113						2

As we can see we needed five comparisons from the DM to fill in the matrix, five comparisons were carried out on the basis of transitivity, but only two of them were confirmed more than once. So, we needed to carry out three more comparisons to check the information (121 with 113; 131 with 112; 131 with 113). If the result is the same we can construct the joint ordinal scale on the basis of the matrix (C). We will not analyze here the independence of criteria just assuming it - the elements of the matrix near the second reference situation which are to be the same in such case are shown in matrix (D).

Matrix D

	233	133	323	313	332	331
233	2	3	1	?	1	?
133		2	?	?	?	?
323			2	3	3	?
313				2	?	?
332					2	3
331						2

According to the matrix C we can rank order vectors of the list L_1 as follows: 211, 112, 121, 131, 311, 113. Using this scale, let us try to compare the initial three alternatives: the first value upon any criterion has the first rank; the second value upon the first criterion (211) has the second rank, the second value upon the third criterion (112) has the third rank and so on.

Let us change in each vector, describing real proposals the numbers of values upon criteria for ranks in the joint ordinal scale. The result is represented in the third column of table 3.3.

Table 3.3. Vectors for job proposals

Job proposals	Initial vector	Rank vector
Proposal 1	123	147
Proposal 2	231	251
Proposal 3	312	613

The newly obtained rank vectors differ from the initial ones in that in them we are able to compare values not only for one criterion but also across the criteria. Using this property these vectors may be ordered as follows: Proposal 2 is more preferable than Proposal 1 and Proposal3; Proposal 3 is more preferable than Proposal 1. The explanations for these conclusions are presented in Figure 3.4.

The evaluation of the procedure's complexity for the decision maker

The proposed procedures for construction and implementation of the joint ordinal scale for rank-ordering multiattribute alternatives have a number of advantages:

- use of simple and understandable information (judgments) from a decision maker;

- provision of thorough consistency check for the involved assumptions (transitivity and independence);

- easy explanations of the results;

- theoretical validation.

Figure 3.4. Explanations for comparison of job proposals

1) Proposal 2 is more preferable than Proposal 1 as according to the JOS
    value 2 criterion 1 IS MORE PREFERABLE than value 2 criterion 2
    value 3 criterion 2 IS MORE PREFERABLE than value 3 criterion 3
    value 1 criterion 3 IS EQUAL TO value 1 criterion 1

2) Proposal 2 is more preferable than Proposal 3 as according to the JOS
    value 2 criterion 1 IS MORE PREFERABLE than value 2 criterion 3
    value 3 criterion 2 IS MORE PREFERABLE than value 3 criterion 1
    value 1 criterion 3 IS EQUAL TO value 1 criterion 2

3) Proposal 3 is more preferable than Proposal 2 as according to the JOS
    value 3 criterion 1 IS MORE PREFERABLE than value 3 criterion 3
    value 1 criterion 2 IS EQUAL TO value 1 criterion 1
    value 2 criterion 3 IS MORE PREFERABLE than value 2 criterion 2

At the same time there are certain problems with its implementation in real tasks. First it requires rather many responses from the decision maker. While comparing vectors near reference situations, we need to fulfill a matrix of pairwise comparisons of the size $N_1 \times N_1$, where $N_1 = \sum_{q=1}^{Q}(n_q - 1)$. Let estimate the number of required responses from the decision maker in the worst case for Q criteria with $n_q = m$ for any $q \in K$.

In general we are to fill in $Q \times (Q-1)/2$ of sub matrices for each two criteria of the size $(m-1) \times (m-1)$ (that is the over diagonal part of the initial matrix.) So, initially we have to fill in $Q \times (Q-1) \times (m-1)^2$ elements.

If we ask the DM to compare vectors y_1 and y_2 from L_1 with $y_{1s} = x_{si} \neq x_{s1}$ and $y_{2t} = x_{ti} \neq x_{t1}$ (i<m), then if according to the decision maker's opinion $(y_1, y_2) \in P_{DM}$ or $(y_2, y_1) \in P_{DM}$, using the relation P^0, we are able to conclude that this relation is maintained for vectors from L_1 with x_{sj} in the first case, and x_{tj} in the second case, where j>i. Thus if i=2, we are able to fill in (m-1) elements in the matrix in the case of strict preference in the pair of vectors from L_1. If the DM responds that $(y_1, y_2) \in I_{DM}$, we are able to define the relation for vectors with x_{sj} and x_{tj}, j>i. Therefore, in any case by one respond we are able to fill in m-i+1 element (i=1,2,...,m). Asking the DM to compare vectors corresponding to the diagonal elements of a sub matrix (for two criteria), the number of questions to the DM will be equal to $(m-1)^2 - (m-1) \times (m-2)/2$. For the whole matrix the number of questions to the DM in the worst case will be equal to

$$N_r = 0.5Q(Q-1)[(m-1)^2 - (m-1)(m-2)/2] = 0.25Q(Q-1)m(m-1).$$

This leads us to the following numbers: Q=5, m=3, N_r =30; Q=6, m=3, N_r =36; Q=5, m=4, N_r =60.

We must realize that in real tasks this number may be less, as relations of indifference give more information and the transitive closure may fill in elements of other sub matrices. But if we consider additional questions in the case of contradictory answers, and the construction of the JOS near the second reference situation, this number may become rather large, though explainable and the questions do not require much time for consideration.

The interactive system ZAPROS

ZAPROS is a Decision Support System (DSS) for rank ordering of multiattribute alternatives.

ZAPROS is a system which helps a decision maker (user) to rank order a large number of alternatives estimated upon a set of criteria. The system supports three main tasks throughout the process of task solution:

- entering/editing initial data (alternatives, criteria and their scales, estimation of alternatives against criteria);

- construction of the decision rules for the comparison of multiattribute alternatives on the basis of the decision-maker's (user's) preferences (through a flexible dialogue with the user);

- implementation of the elicited rules for pairwise comparison of real alternatives and their rank ordering.

The decision support system ZAPROS is a user-friendly tool. The system allows to use natural-like language for evaluation of multiattribute alternatives (through implementation of verbal estimations of criterion values and qualitative comparison of specially formed alternatives). The system gives the opportunity to detect and eliminate possible errors in the user's responses, verify his (or her) preference system for independence conditions, provide explanations of the result. The dialogue is organized on the basis of psychologically valid procedures.

The system can be used in many real situations, where it is necessary to select one or a group of the more preferable alternatives out of their large set or to rank order the alternatives: job selection; selection of a house to buy; evaluation of projects or applicants and so on.

First, the system of criteria is elaborated. This process is not formalized and is carried out prior to the computer use.

Second, the decision rule construction is carried out according to the method ZAPROS described earlier.

Once the decision rule (Joint Ordinal Scale in our case) is constructed and all alternatives are estimated against criteria, the system carries out the rank ordering of alternatives in accordance with the constructed decision rule. Main steps of the process are presented at figure 3.5.

Elicitation of the decision maker's preferences

In accordance with the earlier described method, decision maker's preferences are elicited through comparison of pairs of vectors differing in components upon only two criteria. The process is carried out near two reference situations which are combinations of all the most preferable or the least preferable criterion values correspondingly.

A list of all hypothetical alternatives from L_1 is formed. Information about the decision maker's preferences is obtained by presenting the decision maker with pairs of vectors from L_1 for comparison.

Figure 3.5. Main steps in rank ordering multiattribute alternatives

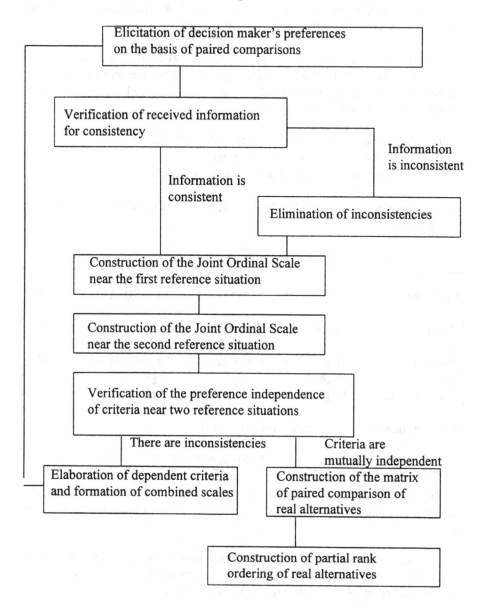

The task is to fill in matrix of the size $N_1 x N_1$ with results of pairwise comparisons of alternatives from L_1, where $N_1=|L_1|$. All questions, necessary to compare all vectors from the list L_1, must be asked.

First, information about the comparison of vectors from L_1 according to the dominance relation P^0 is filled in (without presenting these pairs to the decision maker). Then the decision maker is to compare the displayed vectors, using the following variants of possible answers:

1) alternative A is more preferable than alternative B;
2) alternatives A and B are equally preferable;
3) alternative B is more preferable than alternative A.

An example of such a question for criteria presented in Table 3.1 is presented in Figure 3.6.

Figure 3.6. Visualization of hypothetical alternatives for comparison

CRITERIA	ALTERNATIVE A (2111)		ALTERNATIVE B (1211)
Originality idea	There are new elements		Absolutely new
Prospects	High probability of success.		Success is rather probable.
Qualification	High	=	High
Level of work	High	=	High
Possible answers:	1. A is preferred to B		
	2. A and B are indifferent		
	3. B is preferred to A.		

The decision maker's response is used to construct a transitive closure of the constructed binary relation. The interview is carried out till the matrix is filled in.

As errors are possible (in decision maker's responses), several additional comparisons are carried out to guarantee the requirement for the double check of all comparisons obtained through transitive closure.

If the decision maker's response does not coincide with the one in the matrix, the error message is displayed, and the triple of contradictory responses with explanations are presented (see figure 3.7).

The decision maker is to analyze the situation and to mark what response (or responses) are to be changed to resolve the situation.

Figure 3.7. Visualization of decision maker's contradictory responses and explanations

CRITERIA	Alternative A	Alternative B	Alternative C
Originality	Absolutely new idea	There are new elements	Absolutely new idea
Prospects	High	High	High
Qualification	High	High	Unknown
Level of work	Low	High	High

$A > B > C$, last response $C > A$

Earlier it was marked that alternative A was preferred to B, alternative B was preferred to C. That's why A was preferred to C. Now you say that alternative C is preferred to B.

What comparison would you like to change ?
Possible answers : AB, BC, AC.

The necessary changes are done in the matrix of pairwise comparisons. It is supposed that we only start the interview with a DM (that is we have only these three answers for pairwise comparisons of A with B; B with C; C with A). And also at this moment we know that these answers do not contradict to each other. Let us include each of these answers to P_{DM} or I_{DM} accordingly and carry out the transitive closure of the obtained relation as in the initial interview with a DM. After that the system carries out further formation of the binary relation on L_1.

During this procedure for pairs of vectors from L_1 for which the relation has not been defined, the information is obtained from the previous responses of a DM. After that the transitive closure is built and so on.

This way guarantees that previous DM's answers do not contradict to the newly built relation (as we use only answers for those pairs of vectors for which previous answers have not predefined some relation). As a result we obtain new transitive relation on the set L_1 in which the necessary changes are being made, but all previous answers not contradictory to the new ones are maintained unchanged.

After that, the condition of "double test" for each pairwise comparison is checked for this new information.

As a result of an interview with a DM and transforming his (her) responses to a non-contradictory variant the transitive and reflexive relation R_{DM} of a linear quasi-order on the set L_1 is built.

Analogous interview with the decision maker for construction of the linear quasi-order on the set L_2 near the second reference is carried out. The example of questions asked is presented in figure 3.8.

Figure 3.8. Visualization of hypothetical alternatives for comparison near the second reference situation

CRITERIA	ALTERNATIVE A (3143)	ALTERNATIVE B (1443)
Originality	Development of previous ideas	Absolutely new idea
Prospects	High probability of success	Success is hardly probable
Qualification	Low	= Low
Level of work	Low	= Low
Possible answers:	A is preferred to B A and B are indifferent B is preferred to A	

Implementation of the information about the decision maker's preferences

As has been stated above, to use the obtained information effectively the preference independence of criteria is required. The verification of this requirement is carried out by the system on the basis of comparisons carried out near two reference situations. If the criteria are mutually preferentially independent, the Joint Ordinal Scale is used for comparison of real alternatives.

The Joint Ordinal Scale is constructed on the basis of the comparisons carried out near the first reference situation as follows. On the set L_1 the linear quasi-order R_{DM} is built. Let us assign to each vector from L_1 the number of it's place in the resulting rank ordering. This number will be called the *rank* of the respective vector. It is clear that equal ranks will be assigned only to equally preferable vectors (Furems & Moshkovich, 1984).

As each vector from L_1 has only one component different from the most preferable, the rank assigned to the vector is considered to be the rank of this different component (example of such a scale is presented in figure 3.9).

The construction of the Joint Ordinal Scale leads to a simple and understandable rule for comparison of multiattribute alternatives. Let us denote as $r(x_{jq})$ the rank in the joint ordinal scale of the level j upon criterion q (this means that the hypothetical alternative from the list L_1 with the best levels upon all criteria, but q and with x_{jq} upon the q-th criterion, has this rank in the order of alternatives from the list L_1).

Let us consider that the smaller rank identifies better criterion level. All best levels of each criterion have the same rank which is equal to one .

Let us now have two alternatives y_i and y_j for comparison: $y_i=(y_{i1},y_{i2},...,y_{iQ})$ and $y_j=(y_{j1},y_{j2},...,y_{jQ})$. For each component y_{iq} of alternative y_i we are able to assign r_{iq}, which is the rank of the corresponding criterion level in the Joint Ordinal Scale. In the obtained rank vector let us rearrange components in a non-descending order. Now $r_{i1} \leq r_{i2} \leq ... \leq r_{iQ}$.

Analogously we obtain vector r_j for alternative y_j by it's rank in the joint ordinal scale. Thus, we have $r_i = (r_{i1},r_{i2},...,r_{iQ})$ and $r_j = (r_{j1},r_{j2},...,r_{jQ})$.

Rule. Alternative y_i is not less preferable than alternative y_j if $r_{iq} \leq r_{jq}$ for each q from K (Larichev & Moshkovich, 1991, 1995).

The correctness of such a rule is almost evident if to recall that to state y_i to be not less preferable than y_j it is enough to find for each component of y_i not more preferable (according to the Joint Ordinal Scale) component of y_j. Introduced ranks reflect the preference for different criterion levels according to the Joint Ordinal Scale. Thus, the smaller the rank, the more preferable this criterion level is. If we rearrange ranks in a non-descending order, than the first rank marks the most preferable criterion level in respective alternatives. If the most preferable component of alternative y_j has smaller rank than the most preferable component of alternative y_i, it is clear than alternative y_i can be preferred to alternative y_j, and so on.

Figure 3.9. Joint Ordinal Scale (ranking of vectors from list L_1)

RANK	JOINT ORDINAL SCALE (ordered values)	Vector
1	Absolutely new idea and/or approach	1111
1	High probability of success	1111
1	Qualification of the applicant is high	1111
1	The proposed work is of high level	1111
2	The proposed work is of middle level	1112
3	There are new elements in the proposal	2111
4	The proposed work is of low level	1113
5	Qualification of the applicant is normal	1121
6	Further development of previous ideas	3111
7	Qualification of the applicant is unknown	1131
8	Success is rather probable	1211
9	Qualification of the applicant is low	1141
10	There is some possibility of success	1311
11	Success is hardly probable	1411

If there are discrepancies of comparisons carried out upon Joint Ordinal Scales built near two reference situations, the system informs the decision maker about the situations. It is supposed that the consultant together with the decision maker are able to carry out context analysis to reveal the dependent criterion (criteria). In this case it is proposed to combine dependent criteria and repeat the preference elicitation process for the modified system of criteria.

Rank-ordering of real alternatives

Using Joint Ordinal Scale it is possible to compare pairs of real alternatives.

Some pairs may not be compared. As a result, some partial (not connected) rank order on the set of alternatives A is constructed.

This leads us to the question of how to rank-order alternatives on the basis of the matrix of pairwise comparisons of real alternatives. There are several approaches to this problem. The system provides possibilities to rank order alternatives upon basic principles, adopted by specialists in data analysis. The most popular four of them are the following.

1. Sequential selection of non-dominated alternatives. The idea is to select a group of non-dominated alternatives (containing the best one). After that these alternatives are given rank 1 and excluded from the set. Once again a set of non-dominated alternatives out of this subset is selected. These alternatives are considered to be of the second rank. And so on up to end of the initial set of alternatives.

2. Sequential selection of non-dominating alternatives. The second principle works in a reverse manner. First, alternatives, which do not dominate any other alternatives are selected. This subset is considered to be the least preferable one. After that these alternatives are excluded from the initial set of alternatives and the following subset of non-dominating alternatives is selected. After the subset becomes empty, the ranks of the alternatives are recalculated in a reverse order (the first selected have the greatest rank). It is done to have the smaller ranks for better alternatives.

3. Sequential selection of alternatives, which dominate maximum of other alternatives. For each alternative the number of dominated ones is calculated. The smallest ranks are assessed to alternatives dominating the maximum number of other alternatives.

4. Sequential selection of alternatives, which are dominated by minimum of other alternatives. For each alternative the number of alternatives which dominate it is calculated. The smallest ranks are assigned to alternatives dominated by the smallest number of other alternatives.

These four principles of ranking of alternatives on the basis of partial information about their pairwise comparisons represent the basic ideas in this field and may be easily applied to the same matrix of pairwise comparisons.

In addition the system is able to carry out alternatives' ranking according to all four principles with calculation of the average rank for the alternative. This information may be used for the analysis of the alternative's place in the ranking.

The decision maker is supposed to select the relevant principle on the basis of the requirements of the task under consideration (see figure 3.10).

Figure 3.10. Menu of possible principles for rank ordering of real alternatives

---

System is able to rank order alternatives according the following principles:

1. Sequential selection of non-dominated alternatives.
2. Sequential selection of non-dominating alternatives.
3. Sequential selection of alternatives, which dominate maximum of other alternatives.
4. Sequential selection of alternatives, which are dominated by minimum of other alternatives.
5. Rank ordering of alternatives according to four principles with calculation of the average rank for each alternative.

YOUR SELECTION:

The resulting rank ordering may be presented as in figure 3.11.

Explanations

There is an easy possibility to get explanations for comparison of any two alternatives in the built ranking according to the presented above rule (see figure 3.12). If these alternatives are incomparable on the basis of this rule, then the resulting relation between them is explained with the help of additional alternatives.

Let alternative y_i has smaller rank than alternative y_j in the final ranking. At the same time on the basis of the JOS alternatives y_i and y_j are incomparable. Then if the ranking was done according to the principle of sequential selection of non-dominated or non-dominating alternatives, the alternative y_k is searched for, which is incomparable with alternative y_i but is more preferable than alternative y_j.

Figure 3.11. Rank ordering of real alternatives

Rank ordering of alternatives (the principle of <u>Non-dominated alternatives</u>).

　Rank 1 is assigned to 7 alternatives
　(Numbers are: 1,11,15,3,7,10,17)

　Rank 2 is assigned to 5 alternatives
　(Numbers are: 2,4,16,5,8)

　Rank 3 is assigned to 6 alternatives
　(Numbers are: 6,12,13,18,19,20)

　Rank 4 is assined to 2 alternatives
　(Numbers are: 9,14)

If the ordering is according to the number of dominated (dominating) alternatives, then the alternative y_k is searched for, which is dominated by y_i but is not dominated by y_j.

Example of possible messages for explanation of the results of comparison of real alternatives when they are incomparable upon the JOS is given in figure 3.13.

Conclusion

The described method and system aim the construction of a rank-ordering of the set of alternatives on the basis of a decision-maker's preferences. It is focused on elicitation of the DM's preferences in a qualitative (ordinal) form and on the implementation of a logical transition to a decision rule for comparison of real alternatives. The criteria with verbal scales are used. There is a possibility to organize a reasonable interview with a DM to obtain information about his (her) preference system, to detect possible errors in his (her) responses and to correct them on the basis of their analysis in a dialogue with a DM. Possibilities for the analysis of criteria dependency are shown. This allows to use effectively the obtained information. The DM is also able to decide on the necessity of this analysis and to understand the compromise between the validity of the result and the efforts required.

Figure 3.12. Example of explanations using comparisons upon JOS

Alt. 6 (4122) IS MORE PREFERABLE THAN Alt. 9 (3343)
because as a result of the interview it is stated that:
value 4 upon criterion 1 (alt.6)   IS MORE PREFERABLE THAN value 3 upon criterion 2 (alt.9);
value 1 upon criterion 2 (alt.6)   IS MORE PREFERABLE THAN value 3 upon criterion 2 (alt.9);
value 2 upon criterion 3 (alt.6)   IS MORE PREFERABLE THAN value 4 upon criterion 3 (alt.9);
value 2 upon criterion 4 (alt.6)   IS MORE PREFERABLE THAN value 4 upon criterion 4 (alt.9).

Figure 3.13. Example of an explanation in case comparison upon the principle of ranking

Alt. 2(3311) IS MORE PREFERABLE THAN Alt. 12 (1243)
because:
they are incomparable on the basis of JOS
but the least preferable alt. 1, dominating alternative 2
HAS SMALLER RANK THAN
the least preferable alternative 4, dominating alternative 12.

The system ZAPROS allows a user friendly interface and does not require special computer skills or the knowledge in decision making field.. At the same time, the method ZAPROS-LM itself requires adequate knowledge in the area of multicriteria decision making. That's why the system can be used by the consultants working with the decision makers .

An application of the method ZAPROS for R&D evaluation

Among various problems of R&D planning there is one most often faced (Zuev et al., 1979; Larichev et al., 1978; Larichev, 1982; Larichev & Moshkovich, 1996), notably, selecting group of better R&D proposals out of a large set of applications. This type of work was carried out by the planning body of the Russian Academy of Sciences in Soviet times. It is one of the main duties of the Russian Basic Research Fund (RBRF) in nowadays. This is the task solved by numerous foundations promoting research in different countries (National Science Foundation, Soros Funds...).

The usual way of solving the problem (though they may be different routines in different organizations) is to use experts to evaluate each application and to make decisions on the basis of these evaluations.

The Soviet Academy of Sciences authorities were not entirely satisfied with the state of the things. First, they had the impression that they delegated the decision making to experts (which could be influenced by subjective factors), and did not know what to do when the opinions as to accepting or rejecting the proposal were opposite. The resulting decisions could be heavily influenced by precedents: e.g., the money were given mostly to those who had already made a name. In that case young promising researchers could be overlooked. In other cases special areas of research were given more support than the others due to the interests of higher officials in planning committees or influential experts. In addition, as there could be hundreds or thousands of proposals (which all had to be analyzed and taken into account in the selection process), the processing time and making decisions took too much time, making some financial decisions too late in a year.

As a result, it was decided to elaborate a special effective decision making procedure with explicitly stated policies for evaluation and selection of R&D projects.

Analysis of the task revealed several peculiarities, which made it reasonable to use the method ZAPROS for it's solution:

- There is a large number of diversified alternatives, which it is necessary to partially rank order (to be able to finance better proposals within the resources at hand).

- The diversified nature of alternatives requires to use experts to evaluate them (the decision maker is not an expert in this case).

- The decision maker in this task is the planning body which is interested in expressing it's explicit policy in evaluating proposals, using a set of unified criteria for each alternative. This policy is not based on the proposals at hand but rather on the position of the planning body in it's requirements for R&D projects.

- The policy (decision rules) of the planning body are to be elaborated before the complete set of alternatives is formed, to be able to implement them immediately after the arriving of the last proposal.

- Characteristics (criteria) which are to be used for evaluating of R&D projects are mostly of qualitative nature. It is difficult to use numerical form of their assessment by different experts without the possibility of biases in their estimates (due to individual perception of adequacy between verbal and numerical values for them).

In this task three main interested groups are: the applicants (submitting the R&D projects), the experts (who are to evaluate projects), and the decision maker (planning body) who is to introduce and implement the policy to evaluate and select the appropriate projects.

The first step recommended by consultants was to elaborate system of criteria for evaluation of R&D projects. That set of criteria was to be used by experts while evaluating R&D projects, and the same set was to be used by the decision maker while elaborating the decision rules for selection of better alternatives.

The task under consideration was the formation of a 5-year plan of R&D research in Academy of Sciences. Elaboration of the plan involved contributions from the authors of proposals, the decision maker, and consultants. The information concerning the set of criteria was available to everybody. The decision rule was developed by the consultants and the DM for the later use.

The decision maker expected the consultants to submit explicit verifiable recommendations consistent with his policy. This placed specific constraints on the decision rule elaboration technique. The new plan formation procedure differed from the old one in that the experts were to receive a special questionnaire and the decision maker would take decisions on the basis of the formulated decision rule.

In practice, the number of proposals could ranged from hundreds to several thousands. The R&D projects which were subjected to evaluation largely represented applied research, thus they were mostly oriented towards the solution of the specific problems. The set of criteria used in the evaluation had been modified several times though the stages of the analysis, and as a result was restricted to the following five criteria with qualitative ordinal scales.

CRITERIA FOR R&D PROJECTS' EVALUATION

Criterion A. Scale of R&D project

A1 - R&D project is a part of several programs in the area
A2 - R&D project is a part of one of the programs in the area
A3 - R&D project is not a part of any program in the area

Criterion B. Contribution to the solution of a problem

B1 - As a result of the R&D project the problem would be completely solved
B2 - The solution of the problem will benefit from the R&D project
B3 - R&D project will not essentially benefit the solution of the problem

Criterion C. Versatility of the expected results

C1 - The result of the R&D project will be presented as a working model or pattern
C2 - The result of the R&D project will be presented as a report with recommendations for it's implementation
C3 - The result of the R&D project will be presented as a report describing the principle possibilities for the problem solution

Criterion D. Novelty of expected R&D result

D1 - this is an absolutely new idea in the area

D2 - analogous studies show the perspectives of the research

D3 - analogous results are known

Criterion E. The applicant's qualification

E1 - The experience, qualification and past accomplishments single out the applicant among other researches in the area

E2 - The applicant's experience and qualification are sufficient for the given job

E3 - The applicant's experience and qualification in the area are not known

E4 - The applicant's experience and qualification in the area do not suffice for the given job.

Criterion F. Availability of research background

F1 - The applicant has completed a major portion of the given R&D project. The remaining part of research poses no problems.

F2 - The R&D project activities face a number of problems. There are some ideas concerning their solution and defined lines of research

F3 - The R&D project depends on the solution of a number of difficult problems. There are no ideas concerning their solution.

The description of the situation in terms of qualitative criteria is a verbal decision model. The formulations of scale values reflected those grades the planning body took into account in decision making. In fact, they represented the language of communication between the planning body officials and experts, and a means for obtaining relevant information.

To our mind this approach helps significantly to increase the reliability of information furnished by the experts. The latter tend to be biased to the greatest extent when they are offered the opportunity of evaluating the decision alternatives as a whole and allowed to determine their strengths and weaknesses on an ad hoc basis. In the case where the set of criteria and their possible values are made available at the start, the expert has to consider the assessed objects (R&D projects) from the point of view of the planning body's preferences. In evaluating a project against each criterion, the expert selects one out of several submitted variants as appropriate in characterizing the project. Should he (or she) be biased, and would like to "correct" the actual R&D project's value, the assessments on individual criteria are easy to verify. And for the expert this raises the danger of being considered "professionally incompetent".

It is worth noting that the set of criteria was defined on the basis of the decision maker's desire to emphasize qualities substantial for a comprehensive evaluation of R&D projects. The criterion values were also developed with the decision maker's assistance. Their quantity was determined by the decision maker's intention to single out certain distinct quality levels to be subject to measurement. Each criterion value was thoroughly reviewed in a session with a group of potential experts.

All incoming projects were divided into groups clustered by subject matter. Experts were nominated to evaluate the projects by multiple criteria. First, each alternative was evaluated by one expert and then the estimates were verified by some other competent expert.

Further, method ZAPROS was used to elaborate decision maker's preferences and use them for comparison of real alternatives. In this case the "decision maker" was presented by three officials of the planning body, responsible for making decisions in subject divisions. In the process of analyzing the problem, it was decided that the best way to formulate united decision rule, would be to make comparisons of hypothetical alternatives, presented by the ZAPROS system by all three of them in agreement. Thus, the procedure was used while all of them were present at once, and the responses were entered into the system after they agreed on the appropriate variant (Zuev et al.,1979).

It took three iterations to work out the resulting decision rule (Joint Ordinal Scale). The first variant (which used 8 criteria) revealed preferential dependencies among criteria. They were resolved by combining dependent criteria in one with more complicated scales (combinations of their criterion values). The process restricted the number of criteria to 6, but two of them with rather long scales.

The second iteration (working with 6 criteria, two of them modified and with rather long scales) revealed that these long scales for the modified criteria were not convenient and necessary for preference elicitation. In the process of hypothetical alternatives comparisons, smaller scales with more distinct criterion values were formulated.

The third iteration allowed to construct the Joint Ordinal Scale on the set of these criteria, which satisfied all the decision makers, and preferences were stable near both reference situations. This Joint Ordinal Scale was used to partially rank order 750 R&D projects, resulting in 36 ordered groups of projects (according to the principle of non-dominated alternatives selected by the decision maker).

There was carried out a retrospective examination of the actual results of the R&D projects (Chuev et al., 1983). The results of each accepted R&D project were evaluated by special commissions in terms of a four criteria (with verbal scales):the scale of R and D project's final result;real contribution to the solution of a problem;versatility of result; practical utility.The decision rule have been developed also with the help of ZAPROS method.

The comparison of expected results with real ones have been made for 750 projects. The prognostic power of ZAPROS could be validated from its success in predicting success or failure of one or another project at the stage of planning. To do so, the accepted projects in quasiranking were divided with the help of DM into three groups by their quality. The same was done for the quasiranking of the resulting actual estimates of projects. Analysis demonstrated that on the set of 750 projects the correlation was 82% (Chuev et al., 1983). In our view, this is a good result for a large sample of typical nonstructured problems.

The result supported the idea that such analysis may be useful while making decisions on R&D projects selection. The system (with correspondingly modified sets of criteria and Joint Ordinal Scales) was used for two more 5-year plan formations, providing the planning body with valid information for making decisions. Together with the rank order of projects the evaluations by the criteria were taken into account by planning body officials while making many decisions concerning each project.

Organization systems and decision methods

It is very important, in regard to the applicability of any particular decision method, that the decision makers be ready to apply them. Of course, a more reliable and methodologically validated technique has a greater chance of successful application. The point, however, is not only in the merits or shortcomings of a procedure or a method.

First of all, the new methods and procedures must be adapted to the existing organizational structures and to the traditional ways of gathering and considering the alternatives. Penetrating such systems, the method changes the essence, sharply increasing the rationality and centralization of decision making. At the same time, there is no need for drastic changes of such systems, which are rather difficult for the practitioners.

The problem of applying the new method and procedures is also of a psychological nature. Decision makers tend to share a number of views hampering the improvements in the traditional forms of work. One of them is a consideration implying that a great number of R&D projects (up to several thousands) can well be directly analyzed. It is clear that with complex and different R&D projects such notions are far from realistic. Another notion is that a choice can be avoided either through proportional allocation of resources to all of the options, or by securing additional resources. Experience shows that this unrealistic assumption can result in dissipation of resources. The third notion holds that the application of the new methods and procedures must lead to a reduced decision maker's influence on decision making. Quite the reverse thing occurs with adequate methods. It should be stressed once again that, on the basis of some estimates or other, the final decision is always taken by the decision maker with the account of the existing constraints.

The complexity of problems characterizing R&D planning does not tolerate either an approach which is too simple, or the one with extreme formality. The practical utility of a method depends on it's level of assistance to the decision makers and it's possibility to explain the results.

4 THE METHOD **PACOM** FOR THE SELECTION OF THE BEST ALTERNATIVE

Main ideas of the method PACOM (PAired COMpensation)

Job selection

The task of job selection is wide spread in life. Let assume a department chair in a University is seeking appropriate candidates for an assistant professor position. Usually there may be many (dozens) of applications for each position. Each application is analyzed. As a result 4 applicants are asked for the interview. On the basis of the interviews one of the applicants is selected for the job offer. The department chair decided to work out some rational procedure for their analysis. The first step the department chair decided to do was to elaborate a set of criteria for evaluation of interviewed applicants. As the applicants selected for the interview all had rather good skills in the area of interest, it was enough to use 8 criteria with two possible values for each.

CRITERIA FOR APPLICANTS' EVALUATION

Criterion A. Ability to teach the required courses
A1 - good
A2 - normal

Criterion B. Ability to teach our students
B1 - good
B2 - normal

Criterion C. Evaluation of research and scholarship
C1 - research is of high level
C2 - research is of middle level

Criterion D. Potential for publications
D1 - high
D2 - middle

Criterion E. Potential for leadership
E1 - good
E2 - normal

Criterion F. Match of interests with faculty
F1 - ideal
F2 - to some extent

Criterion G. Communication skills
G1 - very good
G2 - normal

Criterion H. Connections with business
H1 - good connections
H2 - scarce connections

The four interviewed applicants were estimated against the selected criteria. The result is presented in figure 4.1

From this step a special computer program will assist the department chair to carry out the analysis.

Figure4.1. Evaluations of interviewed applicants

Applicant 1	Applicant 2
Ability to teach courses is good	Ability to teach courses is normal
Ability to teach students is good	Ability to teach students is good
The research is of middle level	The research is of high level
Middle potential for publications	Middle potential for publications
Normal potential for leadership	Good potential for leadership
Match of interests is to some extent	Ideal Match of interests
Communication skills are normal good	Communication skills are very
Good connections with business	Scarce connections with business
Applicant 3	**Applicant 4**
Ability to teach courses is good	Ability to teach courses is normal
Ability to teach students is good	Ability to teach students is good
The research is of middle level	The research is of middle level
Middle potential for publications	High potential for publications
Normal potential for leadership	Good potential for leadership
Ideal match of interests	Match of interests is to some extent
Communication skills are normal	Communication skills are ver good
Scarce connections with business	Good connections with business

Let us first select one of these four variants as approximately the best one (if it is not so, we'll correct this in a later analysis). Let consider the applicant 1 as the best choice. Now we are to compare applicant 2 with him. The first thing the system asks you to do is to rank order the disadvantages of each of the applicants relative to one another (see figure 4.2).

Dragging the colored marker and marking in turn, the most important disadvantages for each alternative, we obtain figure 4.2. (disadvantages of each alternatives are rank ordered: 1 is the most important disadvantage).

Figure 4.2.. The resulting ranks for disadvantages of each alternative

Applicant 1	Rank	Applicant 2
Ability to teach courses is good	1	Ability to teach courses is normal
Ability to teach students is good		Ability to teach students is good
The research is of middle level	2	The research is of high level
Middle potential for publications		Middle potential for publications
Normal potential for leadership	4	Good potential for leadership
Match of interests is to some extent	3	Ideal Match of interests
Communication skills are normal	1	Communication skills are very good
Good connections with business	2	Scarce connections with business

Figures from 4.3 to 4.7 illustrate further analysis assisting in making decisions about the alternatives.

Figure 4.3.. Comparison of auxiliary alternatives

Alternative 1-1	Alternative 2-1
Ability to teach courses is good	*Ability to teach courses is normal*
Ability to teach students is good	Ability to teach students is good
The research is of high level	The research is of high level
Middle potential for publications	Middle potential for publications
Good potential for leadership	Good potential for leadership
Ideal match of interests	Ideal match of interests
Communication skills are normal	*Communication skills are very good*
Good connections with business	Good connections with business
Mark the more preferable alternative	Response: 1-1

Figure 4.3. presents not real alternatives but auxiliary ones, which have been formed as follows. Better criterion values of applicants 1 and 2 form the basic variant. Then this basic variant is modified to obtain less preferable criterion values for criteria A and G (most important disadvantages between two alternatives). As a result we obtain alternative 1-1 (with less preferable criterion value of applicant 1) and alternative 2-1 (with the less preferable criterion value of applicant 2).

Let assume the decision maker prefers alternative 1-1 (as marked on the figure 4.3.). The next comparison is presented at figure 4.4..

Let assume the decision maker prefers alternative 1-1 (as marked on the figure 4.3.). The next comparison is presented at figure 4.4..

Figure 4.4.. Comparison of auxiliary alternatives (second iteration)

Alternative 1-2	Alternative 2-1
Ability to teach courses is good	*Ability to teach courses is normal*
Ability to teach students is good	Ability to teach students is good
The research is of middle level	*The research is of high level*
Middle potential for publications	Middle potential for publications
Good potential for leadership	Good potential for leadership
Ideal match of interests	Ideal match of interests
Communication skills are normal	*Communication skills are very good*
Good connections with business	Good connections with business
Mark the more preferable alternative	Response: 2-1

Figure 4.5.. Comparison of auxiliary alternatives (third iteration)

Alternative 1-3	Alternative 2-2
Ability to teach courses is good	Ability to teach courses is good
Ability to teach students is good	Ability to teach students is good
The research is of high level	The research is of high level
Middle potential for publications	Middle potential for publications
Normal potential for leadership	*Good potential for leadership*
Match of interests is to some extent	*Ideal match of interests*
Communication skills are very good	Communication skills are very good
Good connections with business	*Scarce connections with business*
Mark the more preferable alternative	Response: 2-2

As can be easily seen, the alternative 1-2 differs from alternative 1-1 in one component: one more less preferable criterion value of applicant 1 is added to the alternative (criterion C). According to the decision maker's response it is possible to make a conclusion that two disadvantages of the applicant 1 (criteria B and G) are more important for the decision maker that one main disadvantage of the applicant 2 (criterion A).

Next question is presented in figure 4.5.. It is easy to see that alternative 1-3 is obtained from the basic alternative by adding two less preferable criterion values (criteria E and F) of the applicant 1. Alternative 2-2 is obtained from the basic one by adding one less preferable component of applicant 2 (criterion H).

The response indicates that the next disadvantage of the applicant 2 (criterion H) is less important for the decision maker than two disadvantages of the applicant 1 (criteria E and F). On the basis of the decision maker's responses we are able to make a preliminary conclusion that applicant 2 is more preferable for the decision maker than the applicant 1.

To be sure in the correctness of the conclusion it is desirable to check the dependency of the received responses (figures 4.4. and 4.5.) upon the fixed components in auxiliary alternatives (criterion values of the basic variant). To check the dependency the basic variant is being changed to the opposite one: it is formed from the less preferable criterion values of applicants 1 and 2. The corresponding pairs of auxiliary alternatives for comparison are presented in figures 4.6. and 4.7..

Figure 4.6. Comparison of auxiliary alternatives near the second basic variant

Alternative 1-4	Alternative 2-3
Ability to teach courses is good	*Ability to teach courses is normal*
Ability to teach students is good	Ability to teach students is good
The research is of middle level	*The research is of high level*
Middle potential for publications	Middle potential for publications
Normal potential for leadership	Normal potential for leadership
Match of interests is to some extent	Match of interests is to some extent
Communication skills are normal	*Communication skills are very good*
Scarce connections with business	Scarce connections with business
Mark the more preferable alternative	Response: 2-3

Response in figure 4.6. makes it clear that the decision maker's response has not been changed (when substituting better criterion values in fixed components for the less preferable ones). The same is true for responses in figures 4.5. and 4.7..

The decision maker's responses allow us to make the final conclusion that applicant 2 is more preferable for the decision maker than applicant 1 (criterion values for applicants 1 and 2 are presented in figure 4.1).

Figure 4.7.. Comparison of auxiliary alternatives near the second basic variant (second iteration)

Alternative 1-5	Alternative 2-4
Ability to teach courses is normal	Ability to teach courses is normal
Ability to teach students is good	Ability to teach students is good
The research is of middle level	The research is of middle level
Middle potential for publications	Middle potential for publications
Normal potential for leadership	*Good potential for leadership*
Match of interests is to some extent	*Ideal match of interests*
Communication skills are normal	Communication skills are normal
Good connections with business	*Scarce connections with business*
Mark the more preferable alternative	Response: 2-4

Analogously the applicant 2 was compared with the applicant 3. As a result applicant 2 was preferred to applicant 3. Difficulties occurred while comparing applicants 2 and 4. According to the decision maker's responses the main disadvantage of the applicant 2 (middle potential for publications) was less important for the decision maker than two disadvantages of the applicant 4 (middle level of the research and normal communication skills). At the same time the next disadvantage of the applicant 2 (scarce connections with business) appeared to be more important for the decision maker than the disadvantage of the applicant 4 (not ideal match of interests). In circumstances when disadvantages of one alternative can not be compensated by disadvantages of the other alternative, the alternatives are remained incomparable. The conclusion was that applicants 2 and 4 are equally good choice for the department. It is necessary to carry out negotiations with both of them considering their desire to accept the offer and the corresponding conditions. Obtaining additional information about their attitude towards the job offer, it is possible (if necessary) to modify the set of criteria evaluating these two applicants and carry out analogous analysis for selecting the best option.

The carried out analysis illustrates the main ideas of the method PACOM, intended for tasks in which it is necessary to select the best alternative out of their rather small number. If the initial number of alternatives is large (more than 10), it is necessary to reduce the set of alternatives (using more robust methods of selecting a group of better alternatives - see chapters 3 and 5 in this book).

As in other methods described in this book, alternatives are evaluated using verbal descriptions of corresponding quality gradations. The main step in method PACOM is the process of paired comparisons of criterion values. The psychologically valid procedures of preference elicitation are used (rank ordering of relative disadvantages in a pair of real alternatives and comparison of auxiliary alternatives formed on the basis of basic variants).

The aim of comparison of auxiliary alternatives is to compensate disadvantages of one alternative by disadvantages of the other alternative (confirming that disadvantages of the preferred alternative are less important for the decision maker). If the compensation is not found, the alternatives are remained incomparable: this means that it is not possible to compare them on the basis of the formed set of criteria using psychologically valid elicitation processes. Further attempts to compare them may be connected with the attempts to modify the set of criteria used for alternatives' evaluation (see Berkeley et al., 1990). Method PACOM uses admissible operations 011, 013, 023, 032 (see chapter 2).

Task formulation

Let us consider the same multicriteria decision task as in chapter 4 (method ZAPROS-LM), but when there is small number of alternatives to select from (may be 3 or 4 alternatives, but not more than 10). The task is to select the best alternative. This type of tasks often occurs in strategic decision making (Berkeley et al., 1990; Larichev et al., 1994,Larichev et al.,1996).

In general the task formulation may be presented as follows:

Given:
1. $K = \{q_i\}$ i=1,2,...,Q is a set of criteria;

2. n_q is the number of possible values on the scale of the q-th criterion (q$\in$K);

3. $X_q = \{x_{iq}\}$ is a set of values for the q-th criterion (the scale of the q-th criterion); $|X_q| = n_q$ (q ∈ K);

4. $Y = X_1 * X_2 * ... * X_Q$ is a set of vectors $y_i ∈ Y$ of the following type

$y_i = (y_{i1}, y_{i2}, ..., y_{iQ})$; where $y_{iq} ∈ X_q$ and $N = |Y| = \displaystyle\prod_{q=1}^{Q} n_q$;

5. $A = \{a_i\} ⊆ Y$ is a set of vectors, describing the real alternatives.

Required:
to select the most preferred alternative out of the set A on the basis of the decision-maker's preferences.

Let assume as in the previous case that alternatives are evaluated against criteria with verbal ordinal scales. In this case to form the decision rule for comparison of alternatives from the set A in the criteria space is not reasonable, as the number of alternatives is small: on one hand, it is necessary to analyze only a small part of possible combinations of criterion values (corresponding to real alternatives), on the other hand, the decision rule constructed in the criteria space does not guarantee the comparison of real alternatives from the set A.

Taking into account the presented considerations, it was decided to construct the decision support system which is able to carry out direct comparison of real multicriteria alternatives on the basis of ordinal (verbal) information about the decision maker's preferences.

The general approach to the task solution

To solve the problem under consideration method PACOM (PAired COMpensation) has been developed. PACOM is a decision making method for tasks of strategic choice. It allows to structure the choice problem and to provide the necessary analysis and evaluation of possible alternatives to reach the decision.

Method PACOM is based on the assumption that the decision maker has the necessary knowledge about the task he (or she) faces, as well as about the ways to solve it. The method is used to organize and extend the decision maker's ideas about the task and the ways of it's solution, and to assist the decision maker in the choice of the best practically possible alternative.

Choice of the best alternative is carried out on the basis of pairwise comparison of alternatives. Thus, one of the main problems is to have a procedure for comparison of two alternatives, estimated against a set of criteria, on the basis of the decision maker's preferences.

While developing the appropriate procedure main results connected with descriptive data on human capacities to make valid judgments were taken into account (Larichev, 1987; Larichev et al., 1988). That is, first of all the preference for using verbal (ordinal) judgments while eliciting information on decision maker's preferences.

The following assumptions are in the basis of the procedure:

- A decision maker is able to compare the preferability of alternatives' values upon one and the same criterion.

- A decision maker is able to compare the preferability of one of the two alternatives if they differ in values upon only two criteria.

- A decision maker is able to compare the preferability of one of two alternatives if they differ in values upon more than two criteria, but one alternative is more preferable than the other upon all these differing criteria, except one.

A decision maker is able to give one of the following answers:

1. alternative 1 is more preferable than alternative 2;

2. alternative 2 is more preferable than alternative 1;

3. alternatives 1 and 2 are equally preferable;

4. it is difficult to give an answer.

In addition, the assumption about the preferential criteria independence and transitivity of DM's preferences is used (see chapter 3). One of these assumptions is connected with the condition that preferences among alternatives

And transitivity of preferences implies that if alternative 1 is more preferable than alternative 2, and alternative 2 is more preferable than alternative 3, then alternative 1 is more preferable than alternative 3.

Based on these assumptions, a special procedure for comparison of two multiattribute alternatives on the basis of a pairwise compensation principle was elaborated. This principle implies that disadvantages of one alternatives are to be counterbalanced by disadvantages of the other alternative. As a result it is analyzed which of two alternatives has less important disadvantages (or more important advantages).

Procedure for comparison of two alternatives

The goal of this stage in the decision process is to compare a pair of real alternatives on the basis of thorough analysis of the information about them and in accordance with the decision maker's preferences.

Let there be two alternatives a_i and a_j, estimated upon Q criteria. The two corresponding vectors: $a_i=(y_{i1},y_{i2},...,y_{iQ})$ and $a_j=(y_{j1},y_{j2},...,y_{jQ})$. Alternative a_i is more preferable than alternative a_j upon several criteria, and upon the other part of criteria alternative a_j is more preferable than alternative a_i.

Let us consider that alternative a_j is less preferable upon the first m criteria, and alternative a_i is less preferable than alternative a_j upon the rest n (n=Q-m) criteria. $K_1=\{1,2,...,m\}$ and $K_2=\{m+1,m+2,...,Q\}$ are corresponding subsets of criteria numbers. It is evident that $K_1 \cap K_2=\varnothing$

The procedure for eliciting information about the relative preference for different values upon criteria, is based on comparison of basic (hypothetical) alternatives, differing in values upon only two criteria.

Formation of basic alternatives and their comparison

The process of basic alternatives' formation is the following. Let us form an alternative y_i', differing from alternative a_i only by values upon criterion 1 and criterion m+1. For these criteria values are changed by corresponding values of alternative a_j. Let us also form the alternative y_j', differing from alternative a_j by criterion m+1. For these criteria values are changed by corresponding values of alternative a_j. Let us also form the alternative y_j', differing from alternative a_j by estimates upon the same criteria 1 and m+1. Upon these criteria the hypothetical alternative possesses estimates of alternative a_i:

$$y_i'=(y_{j1},y_{i2},...,y_{im},y_{j(m+1)},y_{i(m+2)},...,y_{iQ}),$$

$$y_j'=(y_{i1},y_{j2},...,y_{jm},y_{i(m+1)},y_{j(m+2)},...,y_{jQ}).$$

As a result there are two pairs of alternatives (a_i,y_i') and (a_j,y_j'), in which alternatives differ in estimates upon only two criteria. At the same time upon one of these criteria one alternative is more preferable than the other, and for the other criterion it is vice versa (see figure 4.8.).

In accordance to our assumptions these pairs may be presented to a decision maker for comparison. Let call these pairs of alternatives as *recommended* for comparison.

In the process of comparison of recommended pairs by the decision maker, the results of the decision maker's responses may be presented by the following binary relations. Let us denote:

1. $(y_i, y_j) \in P_{DM}$ if according to a decision maker's opinion y_i is more preferable than y_j;

2. $(y_i, y_j) \in P_{DM}^{-1}$ if according to a DM's opinion y_j is more preferable than y_i;

3. $(y_i, y_j) \in I_{DM}$ if according to a decision maker's opinion y_i is equal to y_j, or if i=j;

4. $(y_i, y_j) \in N_{DM}$ if according to a decision maker's opinion y_i is incomparable with y_j.

As DM's answer: "vector y_i is more preferable than vector y_j", means that $(y_i,y_j) \in P_{DM}$ and $(y_j,y_i) \in P_{DM}^{-1}$, the relation P_{DM} is anti symmetric and anti reflexive.

Figure 4.8.. Comparison of two pairs of hypothetical alternatives in method PACOM

Alternative a_i	Alternative y_i'		Alternative y_i'	Alternative a_j
y_{i1}	y_{i1}		y_{i1}	y_{i1}
y_{i2}	y_{i2}		y_{j2}	y_{j2}
.	.		.	.
.	.		.	.
y_{im}	y_{im}		y_{im}	y_{im}
$y_{i(m+1)}$	$y_{i(m+1)}$		$y_{i(m+1)}$	$y_{j(m+1)}$
$y_{i(m+2)}$	$y_{i(m+2)}$		$y_{j(m+2)}$	$y_{j(m+2)}$
.	.		.	.
.	.		.	.
y_{iQ}	y_{iQ}		y_{jQ}	y_{jQ}

As DM's answer: "vector y_i and vector y_j are equally preferable", means that $(y_i,y_j) \in I_{DM}$ and $(y_j,y_i) \in I_{DM}$, the relation I_{DM} is reflexive and symmetric.

N_{DM} is symmetric but not reflexive.

If P_{DM} and I_{DM} are transitive, then $R=P_{DM} \cup I_{DM}$ is a quasi-order (Mirkin, 1970). The binary relation N_{DM} is intransitive (this means that it is possible that $(y_i,y_j) \in N_{DM}$ and $(y_j,y_k) \in N_{DM}$, but $(y_i,y_k) \notin N_{DM}$.

In general all such pairs are presented to a DM for comparison. As a result a set of partial combinations of estimates (upon small number of criteria, mainly (2-3)) is formed which it is possible to use for comparison of real alternatives a_i and a_j on the base of pairwise compensation rule, analogous to the one introduced in method ZAPROS-LM (see chapter 3).

Definition 4.1. Alternative a_i is *not less preferable* than alternative a_j, if for each criterion s from the subset K_2 there exists criterion t(s) from the subset K_1 such that alternative a_i is not less preferable than vector y_i' , which differs from alternative a_i in components against criteria s and t(s). Against these criteria vector y_i' has criterion values of alternative a_j.

Though this rule resembles the one used in ZAPROS (chapter 4) it is based on different type of information.

Let proof the correctness of this rule in case of preference independence of all pairs of criteria.

<u>Statement 4.1.</u> *If all pairs of criteria from K are mutually preferentially independent, then alternative a_i is not less preferable than alternative a_j if for each $q_s \in K_2$ there exists criterion $q_{t(s)} \in K_1$ such that $(a_i, y_i') \in R$, where*
$y_i' = (y_{j1}, y_{i2}, ..., y_{js}, y_{i(s+1)}, ..., y_{jt(s)}, y_{i(t(s)+1)}, ..., y_{iQ})$,

<u>Proof of statement 4.1.</u> If all criteria are mutually preferentially independent, then according to theorem 3.6 (Keeney & Raiffa, 1976) there exists an additive value function, describing the decision maker's preferences. Thus, the value of alternative a_i can be expressed as

$$v(a_i) = \sum_{q=1}^{Q} v_q(y_{iq}).$$

Analogously, the value of alternative y_i' may ve expressed as

$$v(y_i') = \sum_{q=1}^{Q} v_q(y_{iq}').$$

As all components of alternatives a_i and y_i' are the same except two components, then out of the fact that a_i is not less preferable than y_i', it follows that for each $q_s \in K_2$ there exists $q_{t(s)} \in K_1$ such that

$$k_{q_s} v_{q_s}(y_{iq_s}) + k_{q_{t(s)}} v_{q_{t(s)}}(y_{iq_{t(s)}}) \geq k_{q_s} v_{q_s}(y_{jq_s}) + k_{q_{t(s)}} v_{q_{t(s)}}(v_{jq_{t(s)}})$$

Let add left and right parts of the expression for all $q_s \in K_2$. As a result we get that for alternative a_i combination of values against all criteria from K_2 (plus analogous number of criteria from K_1) is not less preferable than combination of values against the same criteria of the alternative a_j. As according to the statement $|K_1| \geq |K_2|$, it means that for all other criteria alternative a_i values are not less preferable than corresponding values of alternative a_j. As a result, alternative a_i is not less preferable than alternative a_j.

The statement shows that to use the information about comparison of hypothetical alternatives effectively it is necessary to have preferentially independent criteria.

Verification of preferential criteria independence and task modification in case of dependency

Note, that to compare pairs of real alternatives it is necessary to carry out comparisons only for pairs of alternatives of the type (a_i, y_i'). As has been stated above, we also form pairs of the type (a_j, y_j') and present to the decision maker for comparison. This information may be used to check the preference independence of criteria.

Let consider the decision maker carried out the required comparisons. If the relation, given for the pair (y_j', a_i) is the same as for the pair (a_j, y_j'), then it is supposed that the decision maker's preferences for values' combinations $(y_{j1}, y_{j(m+1)})$ and $(y_{i1}, y_{i(m+1)})$ do not depend on the values against the other criteria.

Really, in the first pair the values against all other (than 1 and m+1) criteria were equal to those of the alternative a_i and in the second pair values against all criteria other than 1 and (m+1) were equal to those of the alternative a_j. As a result there is a partial test of the decision maker's preferences for preference independence, and as the level of fixed values represent the level the real alternatives being compared (see figure 4.8.), the test may be considered to be enough representative (as this information will be used for comparison of these two alternatives).

Thus, if the decision maker's responses for two pairs of hypothetical alternatives are the same, the result of comparison is stored and the next two pairs of hypothetical alternatives, differing in values against criteria 2 and m+2 are formed.

If the decision maker's responses are different, a special analysis is carried out. There may be the following underlying reasons for the observed dissimilarity in responses:

1. random factors (tiredness of the decision maker, error, not appropriate attention to the comparison). In this case the decision maker is proposed to analyze the situation and change one of both of his (her) previous responses);

2. alternatives in the presented pairs are too close in the overall value, but the decision maker does not use the equality binary relation. In this case the decision maker is proposed to use this relation in comparison of both pairs;

3. there exits a criterion (or criteria), which different values in alternatives a_i and a_j have influenced the differences in comparisons of both pairs of alternatives (preferential dependence). In this case the decision maker analyzes the situation and marks the influential criterion value. In accordance with the general approach of verbal decision analysis (see chapter 2), the group of dependent criteria is found out (as a rule this groups consists of no more than three criteria) and the group is substituted by one combined criterion and one past criterion with (ordinal) scales for possible values. Alternatives a_i and a_j are evaluated against this modified set of criteria, and this set of (preferentially independent) criteria is used for further analysis.

Described procedure allows to present the task with the help of preferentially independent criteria.

As a result we obtain a set of compared "pieces" of criterion values, which can be used for comparison of real alternatives a_i and a_j on the basis of the compensation rule introduced earlier.

The procedure for comparison of real alternatives on the basis of the obtained information

The resulting binary relation of quasi-order R on the set of criterion values of alternatives a_i and a_j contains the pairs of values' combinations for the part of criteria for which the preference or equality relation has been established. This relation may be used for comparison of alternatives a_i and a_j on the basis of the compensation rule (see definition 4.1).

To use the constructed binary relation for comparison of real alternatives, it is enough to solve the assignment problem on the basis of matrix B of the dimension $M_1 x M_2$, where $M_1 = |K_1|$, $M_2 = |K_2|$. It is evident that a_i may be found out to be not less preferable than a_j only if $M_1 \geq M_2$.(Gnedenko et al., 1986). Each matrix element $b_{kl} = 1$, if $(y_{ik} y_{il}, y_{jk} y_{jl}) \in R$, and $b_{kl} = 0$ otherwise.

If the assignment problem represented in this matrix has a solution, then a_i is not less preferable (more preferable or equal) than a_j.

The presented procedure is theoretically correct, but may be time and labor consuming one due to two main disadvantages:

1. number of pairs recommended for comparison by the decision maker may be rather large (and with no guarantee that as a result of these endeavors real alternatives would be compared);

2. if alternative a_i is less preferable than alternative a_j upon only one criterion but this disadvantage is so important that there is no disadvantage of the alternative a_j which is as important, the alternatives will be left incomparable.

To overcome these drawbacks an additional procedure is proposed. It allows to form hypothetical alternatives for comparison in an iterative goal-seeking mode, quickly establishing the principal possibility for comparison for two real alternatives.

Let K_1 be the subset of criteria against which alternative a_i is more preferable than alternative a_j, K_2 is the subset of criteria against which alternative a_j is more preferable than alternative a_i, and $|K_1| \geq |K_2|$.

The decision maker is asked to rank order the disadvantages of each alternatives (this means to mark against which criteria the differences in values are more important ones). This is a psychologically valid operation (see Larichev, 1984; Larichev et al., 1987). Thus, the decision maker is to rank order disadvantages of alternative a_i (criteria from the subset K_2) , and disadvantages of alternative a_j (criteria from the subset K_1).

Without restricting the task conditions, let assume that for a_i the rank order is: m+1, m+2, ..., Q, and for a_j the rank order is: 1,2,...,m (these are numbers of corresponding criteria). This means that the mostly essential disadvantage of alternative a_i in comparison to a_j is connected with criterion m+1, the next essential disadvantage is connected with criterion m+2, and so on. Analogously, the most essential disadvantage of alternative a_j in comparison with alternative a_i is connected with criterion 1, the next one is connected with criterion 2, and so on.

As our task is to "compensate" each alternative's a_i disadvantage by the corresponding disadvantage of alternative a_j, it is proposed to start the process with the most essential disadvantage of the alternative a_i (because if it is not possible to compensate this disadvantage by any disadvantages of the alternative a_j the alternatives will be left incomparable).

First, we try to compensate the most essential disadvantage of alternative a_i (criterion m+1 in our case) by the most essential disadvantage of the alternative a_j (criterion 1 in our case). Thus, we form hypothetical alternatives y_i' and y_j', in which values against criteria 1 and m+1 are changed (as has been previously described).

If according to the decision maker a_i is more preferable or equal to y_i', then we consider that the disadvantage of the alternative a_i upon criterion m+1 may be compensated by disadvantage of alternative a_j upon criterion 1. In this case we continue with the next compensation (disadvantage of alternative a_i upon criterion m+2 and disadvantage of alternative a_j upon criterion 2).

Otherwise, we consider this disadvantage of alternative a_i to be so important to the decision maker that it may not be compensated by any one disadvantage of alternative a_j. If $|K_1|=|K_2|$ it is possible to make a conclusion that it is not possible to establish that alternative a_i is not less preferable than alternative a_j. In this case it is possible to try to find out if alternative a_j is not less preferable than alternative a_i.

In case $|K_1|>|K_2|$ it is possible to try to compensate one disadvantage of alternative a_i (criterion m+1) with two (or more if necessary) disadvantages of alternative a_j. This means that we form alternative y_i', in which values of the alternative a_i are changed for those of the alternative a_j against criteria 1, 2 and m+1. As a result we have two alternatives for comparison, differing upon three criteria, alternative a_i is more preferable than alternative y_i' against two criteria (1 and 2) and alternative y_i' is more preferable than alternative a_i against one criterion (m+1):

$$a_i=(y_{i1},y_{i2},..., y_{iQ}), \qquad\qquad y_i'=(y_{j1},y_{j2},...,y_{im},y_{j(m+1)},y_{i(m+2)},...,y_{iQ}).$$

If a_i is more preferable or equal to y_i', the compensation is done, and it is possible to start next compensation. Otherwise, we add one more disadvantage from a_j to y_i', and so on. The procedure stops if a_i becomes more preferable than y_i', or if the list of disadvantages of a_j not used for compensation of the disadvantage of the alternative a_i becomes smaller than the number of disadvantages of the alternative a_i "not compensated" yet with disadvantages of the alternative a_j.

If the alternatives are incomparable, it is possible to use the same procedure, trying to establish if alternative a_j is not less preferable than alternative a_j. If it is also not possible, the alternatives are left incomparable. The main steps in the proposed procedure are presented in figure 4.9.

The proposed procedure does not limit the process of compensation to pairs of criteria. At the same time the simplicity for comparison is provided by the fact than one alternative is more preferable or equal to the other one upon all criteria but one. The earlier introduced compensation rule may be reformulated in a more general form.

Definition 4.2. Alternative a_i is not less preferable than alternative a_j if there exits such partition of criteria from the set K that for each subset of this partition, the combination of criterion values of alternative a_i is not less preferable than the combination of criterion values of alternative a_j.

The case of incomparable alternatives

If alternatives a_i and a_j are left incomparable, it is concluded that it is not possible to compare these alternatives using current set of criteria. If the decision maker considers it necessary to compare each pair of real alternatives, then the set of criteria is needed to be analyzed (why preferences expressed against this set of criteria do no allow to compare alternatives).

Figure 4.9.. Main steps in the procedure of comparison of two alternatives

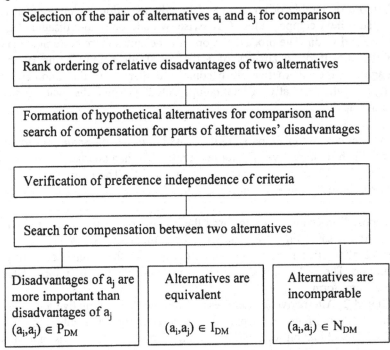

Selection of the pair of alternatives a_i and a_j for comparison

Rank ordering of relative disadvantages of two alternatives

Formation of hypothetical alternatives for comparison and search of compensation for parts of alternatives' disadvantages

Verification of preference independence of criteria

Search for compensation between two alternatives

Disadvantages of a_j are more important than disadvantages of a_i $(a_i,a_j) \in P_{DM}$	Alternatives are equivalent $(a_i,a_j) \in I_{DM}$	Alternatives are incomparable $(a_i,a_j) \in N_{DM}$

It is better to postpone the stage of such analysis till all pairs of alternatives are tried to be compared through the proposed quick procedure, in which case it would be possible to carry out more complete analysis.

That's why it is proposed to leave two alternatives incomparable, but to try to receive additional information about the decision maker's preferences concerning comparison of these alternatives.

The decision maker is asked to mark the minimal changes (Berkeley et al.,1990) in values of one of these two alternatives (potentially more preferable one, e.g. for which $|K_1>|K_2|$ in way to be able to say that one alternative is more preferable or equal to the other. Alternative obtained in such a way is called "adjusted alternative" and is stored in a special list of adjusted alternatives.

Aggregation of information about comparison of pairs of alternatives

When all pairs of real alternatives from the initial list are analyzed using the proposed procedure the general analysis of the results is carried out.

As the aim of the analysis is to select the best alternative if two alternatives are compared the least preferable one is excluded from the list and is not used in further analysis (as it can not be a candidate for the final choice). The more preferable alternative is then compared with the next alternative, and so on. As a result one of the following situations is possible:

1. There is only one alternatives left in the initial list. In this case this alternative is the best choice and the task is fulfilled.

2. There are more than one alternative in the list, but all these alternatives were considered to be equally preferable (this means they are comparable but equal in quality for the decision maker). In this case all these alternatives may not be differentiated using current set of criteria. Any one of them may be considered as a best choice. To be able to choose more preferable alternative out of this set it is necessary to change the set of criteria and carry out the analysis for newly evaluated alternatives.

3. There are several alternatives in the list incomparable between themselves. In this case it is concluded that it is likely there is no enough satisfactory alternative among those in the initial list to make real choice. The decision maker is proposed to analyze the list of <u>adjusted alternatives</u> which has been formed in the process of the analysis. It is necessary to analyze the possibility to obtain a real alternative with such characteristics. If there is such an alternative then it is considered to be the best choice. If there are several such alternatives it is possible to carry out analogous analysis for these alternatives. If there are no such alternatives, then the decision is to be postpone and the process of inventing of new alternative with desirable features is to be organized.

The interactive system PACOM

Method PACOM is the operational part of system ASTRIDA (Berkeley et al., 1990). This system was designed for analysis and support in strategic decision tasks.

The system ASTRIDA provides three main functions: 1) structuring of the problem; 2) choice of the best alternative from the initial list of alternatives; 3) elaboration of new alternative which may not be in the initial list of alternatives. The second part and transition to the third one are presented by the method PACOM.

The thoroughly designed means of flexible interface allow to use the natural for the decision maker language for formulating alternatives, elaborating criteria, and evaluating alternatives against these criteria. The method PACOM is used to assist the decision maker in choosing a best alternative. It is based on pairwise comparisons of specially constructed multiattribute alternatives. In the process of alternatives' comparison the decision maker is provided with the means to modify the description of alternatives. The decision maker is able also in the process of analysis to show the necessity for generation of new alternatives based on the results of the analysis of the initial list of them.

The main steps in the system PACOM are:

1. Analysis of the task starts either with formation of a list of alternatives presented in the task, or with elaboration of a set of criteria, which are to be used to evaluate alternatives.

If the user selects to start the analysis with the introduction of a list of alternatives, the user is provided with means to enter their general description. If the user prefers to start the analysis with the formulation of some best ("ideal") variant of the decision, then the work starts with elaboration of a system of criteria, which are to be used to evaluate future alternatives.

In case of repeated task (if e.g., the organization has been engaged already in the analysis of analogous task and the set of criteria has been formulated), the set may be modified at this step.

2. When list of alternatives and a set of criteria are formed, the decision maker is provided with the means to evaluate each alternative against each criterion. As a result each alternative is presented as a set of verbal values against the criteria.

The examples of the dialogue with the system for introduction of alternatives, criteria, and evaluating alternatives against criteria are presented in figures 4.10, 4.11, 4.12, and 4.13 for the task of gas pipeline route selection (Larichev et al., 1994; Andreeva et al., 1995,Larichev et al.,1996).

3. When alternatives are evaluated against the set of criteria and their number is more than 2, the decision maker is asked to select (if it is possible) the potentially best alternative (according to his/her opinion). This alternative will be considered to be a basis for future analysis. If the decision maker does not want to mark such an alternative the system will choose one of the alternatives by itself (based on some formal evaluations.

Figure 4.10. Formation of the list of alternatives

```
        Block for formation of the initial information
                 Formation of ALTERNATIVES
    ─────────────────────────────────────────────────────

                Enter the list and comments

        ALTERNATIVES        COMMENTS
        Sea route       Trespassing of the gas pipeline under the sea
        Land route      Trespassing of the gas pipeline on the land

        PgDn - Exit comment,  PgUp - Enter comment, Esc - exit
```

Figure 4.11. Formation of the set of criteria

Block for formation of the initial information

Formation of CRITERIA

Enter the list and comments

CRITERIA	COMMENTS
Cost	In hard currency
Ecology	Influence on the environment
Accidents	Possibility of heavy incidents
Sequences	Sequences to accidents
Reliability	Difficulty in restoring gas pipeline after an accident
Unknown factors	Level of uncertainty and unknown factors

PgDn - Exit comment, PgUp - Enter comment, Esc - exit

4. If the number of alternatives is larger than 2, at this step the alternative to be compared with the potentially the best one, is selected. It is difficult for the decision maker to compare complex alternatives, to take into account all advantages and disadvantages of alternatives at once. That's why (as in method PACOM), the system uses special procedure to analyze which disadvantage (disadvantages) of one alternative may be compensated by disadvantage (disadvantages) of the other alternative.

Figure 4.12. Evaluation of alternative 1 (sea route) against criteria al evaluations).

Block for formation of the initial information Formation of ALTERNATIVES' VALUES

Alternative 1　SEA ROUTE

[List of criteria]	[List of values]
Cost	More expensive
Ecology	Inessential influence
Accidents	High probability
Sequences	Inessential sequences
Reliability	Complexity of restoring gas pipeline
Unknown factors	There are uncertainty and unknown factors

F1 - Help,　　　　　F10 - exit

Figure 4.13. Evaluation of alternative 2 (land route) against criteria

Block for formation of the initial information Formation of ALTERNATIVES' VALUES

Alternative 2　LAND ROUTE

[List of criteria]	[List of values]
Cost	Less expensive
Ecology	Essential influence
Accidents	Less high probability
Sequences	Serious sequences
Reliability	Quick restoring of gas pipeline
Unknown factors	High level of certainty

F1 - Help,　　　　　F10 - exit

To compare complex alternatives special hypothetical ones are formed (in accordance with the method PACOM). They differ in values against only two criteria.

To carry out an effective procedure for alternatives' comparisons, the decision maker is asked to rank order relative disadvantages of each alternative (see figure 4.14).

Figure 4.14. Rank ordering of relative disadvantages of two alternatives

RANK ORDERING OF RELATIVE DISADVANTAGES OF TWO ALTERNATIVES				
	Alternative 1 SEA ROUTE		Alternative 2 LAND ROUTE	
Criteria		RANK		RANK
Cost	More expensive	1	Less expensive	
Ecology	Inessential influence		Essential influence	1
Accidents	High probability	3	Less high probability	
Sequences	Inessential sequences		Serious sequences	2
Reliability	Complexity of restoring gas pipeline	4	Quick restoring of gas pipeline	
Unknown factors	There are uncertainty and unknown factors	2	High level of certainty	

Enter the criterion numbers in the order from the most important to the least important disadvantage:

F1-Help, F2-Expand, F3-Status, F4-Command, F5-Auto, F7-Menu, F10-continue

5. Using the introduced rank ordering of alternatives' disadvantages the system forms the of alternatives for comparison by the decision maker, and presents them on the screen (see figure 4.15). If the alternatives are difficult for the decision maker to compare, next pair of alternatives is formed.

Figure 4.15. Comparison of specially formed pair of alternatives

---

COMPARISON OF HYPOTHETICAL ALTERNATIVES

Compare these two alternatives and select the more preferable one. Use arrow keys to choose the response variant. Press ENTER to make the choice

Criteria		Alternative 2	Formed alternative
Cost	1	Less expensive	Less expensive
Ecology	2	Essential influence	Essential influence
Accidents	3	Less high probability	Less high probability
Sequences	4	Serious sequences	Serious sequences
Reliability	5	Quick restoring of gas pipeline	Quick restoring of gas pipeline
Unknown factors	6	High level of certainty	High level of certainty

This alternative is more preferable	This alternative is more preferable
Alternatives are	equal
It is difficult to	give an answer

F1-Help, F2-Expand, F3-Status, F4-Command, F5-Auto, F7-Menu, F10-Continue

---

6. This step is used to implement the information about the decision maker's preferences obtained at the previous step, for comparison of the two previously selected real alternatives on the basis of the method of paired compensation. If the comparison is possible, the less preferable alternative is excluded from further analysis, the more preferable alternative is considered potentially the best one.

7. If it is not possible to compare these alternatives on the basis of the obtained information, both alternatives are maintained in the list for further analysis as incomparable. The decision maker is asked to mark the minimal changes in alternatives' criterion values for one of the alternatives to become more preferable than the other. Results of these changes are stored as adjusted

alternatives in a special list of adjusted alternatives (without changes in the list of initial real alternatives).

10. When the list of available for comparison real alternatives is exhausted the system analyzes the results and presents them to the decision maker for consideration. The goal of this stage of the decision analysis is to consider the results obtained during the previous steps of decision analysis from the point of view of the required decision.

If there is one best alternative is chosen, then the task is considered to be solved.If incomparable alternatives are being left in the list, the system analyzes the reason for incomparability of alternatives (on the basis of saved information).

The system suggests to analyze the list of adjusted alternatives formed in the process of analysis (alternatives formed through improvement o some of criterion values of initial alternatives). All alternatives from this list are presented to the decision maker for analysis of the possibility to achieve these characteristics in a real alternative.

If the decision maker considers it possible to have such an alternative in real life, the task is considered to be fulfilled with this newly formed alternative.

If there is no alternative among adjusted ones which the decision maker considers possible to achieve, the conclusion is made that at the moment there is no satisfying alternatives in the list of real ones. It is necessary to redesign the whole task to be able to find an appropriate solution.

The architecture of the system's software of the system PACOM includes:

1. dialog manager providing the interface with the user on the upper level and the choice of functions inside the system while arranging the task formulation;

2. set of procedural modules providing special forms of the interface with the user and applying special algorithms and procedures of the task analysis;

The complex is programmed using C++ programming language.

Implementation of the method PACOM for the analysis of the gas pipeline route selection problem

In Russia, following the opening of the economy and the drive to obtain export income, oil and gas development now must be publicly justified (Osherenko, 1995; Chance & Andreeva, 1995), One of the large proposed projects in this area was the construction of a gas pipeline around or across Baidaratskaya Bay, on the western side of the Yamal Peninsula.

In 1993 it was decided that Yamal gas fields were to be developed. There were several proposals for the ways of construction the gas pipe-line to deliver the gas from the west-central part of the peninsula to the main pipe-line in European Russia. Two of them were selected for future analysis as reasonable ones: the so-called "sea route" variant (496 km, part goes under the sea), and the so-called "land route" variant (654 km, goes to the south through the western part of Yamal Peninsula, then turns to west and crosses the Ural mountains). More details about these variants may be found in Andreeva et al., 1995.

International decision analysis group of consultants took part in this work. The method PACOM was used to try to compare these variants and generate new alternatives (if necessary). The employed technique was not intended to supplant experts or decision makers, but to help those people clarify their choices and utilize their judgment and knowledge more effectively, both to gather information and to select among "terminal" options (Larichev et al., 1995a; Andreeva et al., 1995).

Research/consulting process

Two Russian members of the research team had several meetings with Gazprom decision makers and then spent several days with the personnel of the institution responsible for the feasibility study - a team of engineers and their leader, as well as with the decision makers. Reports of two institutes involved in the preparation of the two variants under consideration were used in these meetings.

From both the literature and discussions with the experts, the following eight considerations were deemed critical to validation of the route decision:

1.   <u>Route length.</u> The sea option is 496 km, some 160 km shorter that the land route, but 68 km of it is across the waters of the bay.

2. <u>Construction conditions.</u> Construction conditions are difficult for both options. The land route must cross a large expanse of permafrost, rivers, and lakes. The sea route requires working during a limited period (2-2.5 months) each summer when the bay is free of ice. The part of sea route that crosses land has the same permafrost conditions and water obstacles as those found along the land route.

3. <u>Available expertise and technology.</u> Russia, as well as other countries, does not have the appropriate technology for sea-bottom construction in an Arctic basin. It does have considerable experience in building pipelines across land in permafrost areas. A Dutch company, Haarema, was initially selected to develop the sea route. Although a final agreement has not yet been reached, Haarema is likely to be designated as the main partner in these operations, within a new joint venture known as "Petergaz".

4. <u>Construction cost.</u> Cost is an important factor, but quite difficult to estimate. In principle, the cost of the land route can be calculated from the cost of labor, equipment, material and transportation. The cost of the sea route involves additional uncertainties, because prior experience in construction under these conditions is lacking. The participation of the foreign firm increases the use of hard currency in all expenditures.

5. <u>Environmental impact.</u> Both options have a potential negative effect on the environment. The land route will cross many ecologically important habitat areas and migration routes, including reindeer pastures, wildlife refuges, hunting lands, and rivers, and lakes famous for their productivity and fishing resources. The sea route, lying father to the west and being shorter, will have comparatively less impact on the terrestrial ecosystems. It may, however, have unexpected impact on the marine ecosystems of the bay during construction.

6. <u>Risk of pipeline ruptures.</u> The Baidaratskaya Bay situation is unique. Specific physical conditions here could contribute to an accident: (a) the shores of the bay are unstable because of nearly year-round permafrost action and sea-ice impact; (b) indications of ice scouring exist on the bottom of the bay; and (c) experts believe that icebergs can enter Baidaratskaya Bay.

7. <u>Consequences of pipeline ruptures.</u> An accident on the land-based pipeline would probably create an explosion and fire. The result would be the complete destruction of the vegetation cover, an upsetting of the

thermal regime of the permafrost, and the possible death of wild animals. The repair work would create further extensive and long-term environmental damage. An under water accident would have much less environmental impact: it would not involve an explosion, the gas does not dissolve in water and is non-toxic, and the ice cover is not solid, thereby allowing gas to escape via cracks into the atmosphere. The appearance of an air-temperature inversion (promoting air pollution built-up) is not a concern in this non populated area, and is not likely to occur.

8. <u>Recovery time after an accident.</u> This factor affects the reliability of gas supply. In the land-route case, the pipeline could be repaired almost immediately. Repair of the underwater pipeline, on the other hand, could take a long time. Work would be limited to 8-10 weeks when the bay is ice-free. Furthermore, the repair would require special equipment and techniques. One suggested engineering solution is to construct one extra line in the pipeline string into which gas could be diverted in the event of the rupture of one or more of the operating lines; in this case what would be involved is the construction of four pipelines instead of three. There are major uncertainties and risks involved here. The sea option with respect to this eighth consideration, therefore, clearly is much worse.

Structured comparison of alternatives

Systematic analysis of options involved the distillation, from the eight considerations discussed above, of only those criteria demonstrating significant and clear differences between the options and focusing on them. Elimination of considerations where the differences between the two alternatives were very uncertain or otherwise undetectable was the first step. For example, one cannot readily measure the difference in construction time because it is highly uncertain for both routes.

The following six criteria (with symbolic abbreviations) were retained; for some, distinctive assessment issues for further research are identified.

1. <u>Cost of construction (C).</u> This includes any effect resulting from construction conditions and available expertise and technology discussed above. Assessment issue: This is one criterion that lends

itself naturally to quantitative estimation. A foreign firm would estimate the cost of crossing Baidaratskaya Bay (C_{sea}) in United States dollars. How does one then estimate the cost of the land route (C_{land}), which would be paid for in rubles? The obvious solution is to convert the cost of the land route to U.S. dollars as well. The prices for the factors of production can be determined because international markets now exist in Russia for equipment, workers, and material. The initial estimates showed that C_{sea} is a little larger, but further research is ongoing.

2. <u>Ecological impact (E)</u>. E_{land} appears greater.

3. <u>Probability of accident (P)</u>. Assessment issue: Existing data on pipeline operations in the Russian North allow risk assessments for the land route. Known cases of gas pipeline accidents in mountain region can add precision to these assessments. On the other hand, risk assessment for the sea route is difficult. Statistics do not exist for the operation of an underwater pipeline under severe Arctic conditions. Because of the unstable shore and the possibility of ice scouring, P_{sea} appears greater.

4. <u>Consequences of an accident (Q)</u>. Q_{land} is clearly worse.

5. <u>Gas supply reliability (R)</u>. This reflects the time needed to recover from the accident. No country, including Russia, has experience in conducting repair work under the sea-route conditions. The sea route would require, there fore, long-term observations and prolonged investment in scientific research. R_{sea} is clearly much worse.

6. <u>Uncertain and unknown factors (U)</u>. Highly uncertain considerations affecting both options were not distinguished, but a major difference in overall uncertainty was treated as a criterion. The Baidaratskaya Bay route, because it is unique, involves many uncertainties and imponderables. U_{sea} is clearly worse than U_{land}.

Figure 4.16 summarizes these preliminary comparisons with the better highlighted for each criterion.

The PACOM analysis was carried out for these two alternatives.

Generating new options

The system was used to compare these two alternatives on the basis of the principle of paired compensation. The analysis showed that the greater uncertainty and lesser reliability of gas supply for the sea-route option resulted in an overall more negative assessment than the unfavorable ecological impact of the land-route option; at the same time negative consequences of an accident for the sea-route option were less than for the land-route option (see figure 4.17).

Figure 4.16. Evaluation of Sea-Route and Land-Route Options, Yamal Gas Pipeline, Russia

Criteria	Baidaratskaya route	Land route
1. Cost	C_{sea}	C_{land}
2. Ecological impact	E_{sea}	E_{land}
3. Probability of an accident 3	P_{sea}	P_{land}
4. Consequences of an accident	Q_{sea}	Q_{land}
5. Reliability of a gas supply	R_{sea}	R_{land}
6. Uncertain and unknown factors	U_{sea}	U_{land}

Figure 4.17. Comparison of Sea-Route and Land-Route Options, Yamal Gas Pipeline through method PACOM

Criteria	Baidaratskaya route		Land route
1. Cost	C_{sea}	$\Leftarrow$	C_{land}
2. Probability of an accident	P_{sea}	$\Rightarrow$	P_{land}
3. Consequences of an accident	Q_{sea}		Q_{land}
4. Ecological impact	E_{sea}		E_{land}
5. Reliability of a gas supply	R_{sea}	$\Leftarrow$	R_{land}
6. Uncertain and unknown factors	U_{sea}		U_{land}

As a result the options were left incomparable. In this case the system ASTRIDA (as has been stated above) proposes the user a direction for developing a new promising option by adjusting the evaluations of one existing alternatives.

The analysis showed that it was impossible to change anything in the land-route option (as the traditional way of construction set all the characteristics with much certainty).

As far as sea-route was concerned, the main interconnected disadvantages of the Baidaratskaya route are a greater uncertainty and larger probability of accident. The possible direction of development of new variants consisted in finding a way to change these characteristics.

Discussion with the experts suggested ways in which the negative comparisons of the sea option could be removed. A new sea-route option was produced through a reduction in the amount of uncertainty as a result of addition of certain technically feasible features to the original sea-route option (Larichev et al., 1995a):

1. To eliminate the influence of seashore instability it is possible to construct special <u>shafts</u> located at a safe distance from the sea and put the pipeline through the shafts. This will incur additional cost to the sea route variant.

2. To avoid the damage of the pipeline from underwater ice, the pipeline can be laid in special trenches at a depth of 1.5-2 meters (deeper than the floor of the bay). It is deeper than provided by the project and causes additional cost.

3. A very rare but a dangerous event in the bay is an iceberg. An iceberg moving in shallow waters may destroy the pipeline even lying in a trench. To avoid this danger it is necessary to have a special observation service and a special ship to drag the piece of iceberg from the bay. This also costs additional money.

These new features of the adjusted variant allow us to say that:

1. the degree of uncertainty for this new sea route variant and land route variant are approximately equal;

2. the probability of an accident for new variant is not very different for new sea route variant and land route variant;

3. given the possibility of additional service for repairing the tubes underwater, the reliability of gas supply is estimated to be not very different for new sea route and land route variant.

Thus, there is no significant difference between new sea route variant and land route variant except cost and ecological impact. The cost of the new sea route will clearly be greater than that of land route, but the environmental destruction will be greater for land route variant. The new sea route variant can also provide an invisible benefit to the oil and gas industry. Now that question is if the interested parties are ready to pay necessary sums for ecological safety of the region.

Thus, the proposed decision-aiding methodology helped not only to clarify difficult decisions, but also to generate a new variant for seemingly intractable problem.

Conclusion

The described method PACOM is oriented on the support of the decision analysis in selection (choice) tasks with small number of complicated alternatives. Task with these characteristics are often faced while making strategic decisions.

While engaged in the analysis of real strategic tasks (Larichev et al., 1995a; Oseredko et al., 1982) it was marked that the decision maker tried to find the dominating alternative (from the point of view of the stakeholders in the task). If that was not possible, the decision maker tried to modify one of the alternatives. In this way the decision maker tried to construct a new alternative, satisfying requirements of all the stakeholders in the task and being the dominating one in quality.

The method and system proposed in this chapter are based on the marked peculiarities of the decision process. Main emphasis in the work is placed on the implementation of qualitative judgments and "soft" methods of information elicitation. The system is oriented on the process of analysis of the real task faced by the decision maker as a process of flexible and supervised dialogue with the decision maker.

5 THE METHOD **ORCLASS** FOR ORDINAL CLASSIFICATION OF MULTIATTRIBUTE ALTERNATIVES

Main ideas of the method ORCLASS (ORdinal CLASSification)

Let us illustrate the ideas of the proposed method on a simple example of the task of the loan applications' evaluation.

Who is worth the loan?

Suppose that a number of businesses in need of financial resources submitted loan applications to a commercial bank (in Russia). The bank CB (the decision maker in our case) decided to elaborate a unified policy for granting loans in the bank. The decision maker decided to use a special consultant to help in this task.

As a result of the study it was necessary to obtain a definite decision rule: in what cases and at what conditions the loans were to be granted, and in what cases the loan applications were to be rejected. The analysis of previous cases of loans not paid in time, the study of bank's way of dealing with clients, interviews with the bank staff and management carried out by the consultant, resulted in a formation of two main goals in bank operations: "Profit" and "Risk aversion". The first goal states not only the bank's desire to return the loan money but increase the bank's capital due to interest rates. The second goal is to try to guarantee the loan return and to decrease the possibility of loosing money in the loan process.

While stating interest rates the bank uses the data on today's financial market. Wanting to lower the risk, the bank acquires information about the financial reputation of the client and decides on the deposit. The best deposit is the one which it is easy to sell quickly (this is called "high liquidity" of the deposit). The worst deposit is the one which it is not that easy to sell. Three main criteria correspond to these goals:

CRITERIA FOR LOAN APPLICATION'S EVALUATION

Criterion A. Loan term
A1 - short-term loan (3 months)
A2 - medium-term loan (6 months)
A3 - long-term loan (one year or more)

Criterion B. Client's credit history
B1 - credit history is good
B2 - credit history is reasonable
B3 - credit history is bad

Criterion C. Deposit liquidity
C1 - high
C2 - intermediate
C3 - low

These criteria were elaborated by the consultant together with the decision maker. For each of the criteria a scale with three possible gradations was formed. Criterion values are arranged from the most to the least preferable one. As it was not known in advance what clients would apply, it was necessary to have a decision rule to use for any client. It is clear that it is more profitable for the bank to grant short-term loan to a client with good credit history and high liquidity of the deposit. It is also clear that one should not grant long-term loan (in inflation environment) to a client with bad history and low liquidity deposit. But what to do in less evident cases?

To construct the decision rule the consultant proposed the decision maker to use the method ORCLASS (ORdinal CLASSification).

Informative questions

It is evident that the general number of all possible combinations of criterion values presenting these three criteria with the scales is equal to 27. All combinations are presented as small cells in figure 5.1. To construct a decision rule it is necessary to divide all these combinations between two classes: I - the loan is granted; II - the loan is rejected. Let us note, that for two combinations out of their set the appropriate classes are known: for the combination of all the best criterion values the class I is appropriate; for combination of all least preferable criterion values class II is an evident decision.

The direct way to construct a decision rule is to present to the decision maker all these combinations of criterion values for classification. In general case this way may be too time and efforts' consuming for the decision maker (considering that the number of combinations of criterion values in some tasks may be equal to thousands). It is more reasonable to present the decision maker with the combinations which when classified, are able to give us the information about the appropriate classification of a number of other combinations from the general set.

Let us use figure 5.1 for illustration. In this figure our task is presented as a set of three tables. In each table all possible combinations of criterion values for criteria A and B are presented. The left table corresponds to the first criterion values against criterion C, the middle table refers to the second value against criterion C, the right table refers to the third values against criterion C. The decision rule is constructed if each cell in all tables contains the corresponding class number (I or II). Let further denote each cell by the corresponding combination of criterion values (e.g., A3B1C2, A2B1C3, and so on).

Let consider that the combination A3B1C2 (left bottom cell in the middle table in figure 5.1) is presented to the decision maker for classification. Let assume the decision maker decided it to be in class I (loan is to be granted). Then all cells in left columns of left and middle tables in figure 5.1 will also belong to class I as they present better options than A3B1C2 (these cells represent combinations A2B1C2, A1B1C2, A3B1C1, and A2B1C1). These combinations are better for the decision maker as values A1 and A2 are more preferable than value A3, and value C1 is more preferable than value C2.

Figure 5.1. Propagation of the information obtained through a decision maker's response on other combinations of criterion values

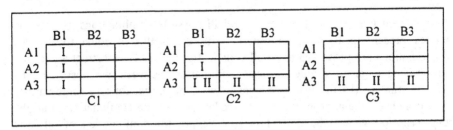

This shows how one classification carried out by the decision maker can give the information sufficient to classify a number of other combinations (4 addition classifications in our example).

Let analyze the consequences of classification of the same combination (A3B1C2) as the second class by the decision maker. In this case it would be possible to classify additionally four other combinations of criterion values as belonging to the second class (combinations A3B2C2, A3B3C2, A3B1C3, and A3B2C3). This would be possible as these combinations have less preferable criterion values against criterion B and/or criterion C.

In advance we do not know what real response of the decision maker for the presented object will be. But if to assume that both variants are equally possible, it is more reasonable to present the decision maker with objects which classification can give maximum additional information (maximum number of additionally classified objects).

Let call the combination of criterion values which classification gives the maximum additionally classified objects the most informative one. To be able to find the most informative combinations of criterion values let introduce the informative index for each combination. This index is calculated for each unclassified combination of criterion values. To calculate it, it is assumed that this combination is presented to the decision maker for classification. Then the number of additionally classified alternatives is calculated for two cases: the decision maker's response is class I, the decision maker's response is class II. In figure 5.2 these indices for all cells in three tables are given. The left number in each cell is for the decision maker's response "class I", and the right number in the cell is for the decision maker's response "class II".

Figure 5.2. Informative indices for all combinations of criterion values

	B1	B2	B3		B1	B2	B3		B1	B2	B3
A1	I	1+17	2+8	A1	1+17	3+11	5+5	A1	2+8	5+5	8+2
A2	1+17	3+11	5+5	A2	3+11	7+7	11+3	A2	5+5	11+3	17+1
A3	2+8	5+5	8+2	A3	5+5	11+3	17+1	A3	8+2	17+1	II
		C1				C2				C3	

Now it is possible to select the most informative combination of criterion values. It is determined upon a set of requirements:

1. it is desirable that both numbers in a cell are similar (the hypothesis about the equal possibility to receive class I or class II as a decision maker's response makes it attractive to receive rather large number of additionally classified combinations for each of the two variants of responses);

2. it is desirable the summed informative index to be the maximum among all cells.

Thus, to choose the cell with the maximum informative index it is necessary and enough:

1. to choose cells in the tables with the smallest difference between the two numbers (in figure 5.2 these cells have equal numbers - the difference is equal to zero);

2. to choose among them the cell with the maximum sum of two corresponding numbers, guaranteeing larger number of additionally classified objects for any decision maker's response (in figure 5.2 it is the cell with numbers 7+7);

3. if there are several such cells, choose any one of them in a random order.

In the beginning of the dialogue with the decision maker the most informative combination of criterion values for our example is cell A2B2C2 (it has numbers 7+7 in figure 5.2, and is marked as X1 in figure 5.3). For any decision maker's response we are sure to be able to classify additionally 7 objects (or cells).

Figure 5.3. Sequence of the most Informative combinations of criterion values

	B1	B2	B3		B1	B2	B3		B1	B2	B3
A1			X10	A1			X4	A1	X11	X8	
	I	I	I		I	I	II		II	II	II
A2			X7	A2		X1		A2	X6	X2	
	I	I	II		I	I	II		II	II	II
A3		X9	X3	A3	X5			A3			
	I	II	II		I	II	II		II	II	II
		C1				C2				C3	

The combination A2B2C2 can be considered as a center of all possible combinations. The decision maker is asked to classify the loan application for *medium-term loan, where an applicant has a reasonable credit history and the deposit has intermediate liquidity.*

Let assume the decision maker considered this application good enough to be granted the loan (class I in this cell). The decision maker's response allows to classify additional seven objects (cells) as belonging to the first class. The number of unclassified objects would be equal to 18. For each of them the informative index is calculated once again (taking into account new information). After that analogously to the previous case the most informative combination is selected. Figure 5.3 shows the sequence of combinations presented to the decision maker to classify each cell in all three tables (X1, X2,...,X11). Thus, it was necessary for the decision maker to classify 11 objects to construct the complete decision rule in the task.

Consistency verification

While receiving judgments from the decision maker it is necessary to check it for consistency. To carry out this check the decision maker must be asked to repeat classifications for the objects, called <u>boundary combinations of criterion values.</u>

Boundary combination of criterion values is such that change of one of it's criterion values leads to the change of the class to which it belongs. Among all cells of the class II for our example (see figure 5.3) it is possible to select boundary combinations (they are cells marked as X4, X7, X9, and X11). These elements define class II for all other combinations belonging to this class. These are combinations which it is necessary to present to the decision maker for the repeated classification to verify the consistency of acquired information.

Analogously, it is possible to select boundary elements for class I (they are X1, X5, and X10), which defined the classification of all other elements in class I. They are also to be once again presented to the decision maker to check the consistency of elaborated decision rule.

Decision rule as a means for explanation of the results

Classifying the presented objects, the decision maker constructs the general decision rule in a step-by-step manner. This decision rule may be described in a verbal way. The rule presented in figure 5.3 may be formulated as follows:

1. For the loan applicant with <u>bad credit history</u>, the loan may be granted only on a short-term basis and with a deposit of high liquidity;

2. for the loan applicant with <u>good credit history</u> even long-term loan may be granted if the deposit is of at least intermediate liquidity;

3. for the loan applicant with <u>reasonable credit history</u> short-term or medium-term loan may be granted if the deposit is of at least intermediate liquidity.

This rule may be used as an explanation for any loan decisions in the bank.

Conclusions from the example

The analyzed simple example shows the main advantages of the method ORCLASS:

1. dialogue with the decision maker is carried out using easily understood language of verbal criterion values;

2. classification is being built through selection of the most informative combinations of criterion values for presentation to the decision maker for classification;

3. it is possible to check the decision maker's responses for consistency;

4. verbal formulation of the resulting decision rule may be used for the explanation of the decisions made.

The multicriteria classification problem in decision making

Along with multicriteria choice problems, people may face multicriteria classification problems. A feature of classification tasks is that it is not necessary to rank order the alternatives, but only to assign them to a small number of decision groups. Quite often these classes (or groups) may be rank-ordered, reflecting different degree of quality. In this sense the alternatives assigned the first decision class are better than those, assigned the second class, etc.

Examples of such tasks may be found in different areas of human decision making: an R&D program leader, who decides which projects to incorporate into the program on the basis of their characteristics; a physician, who determines the severity of the disease on the basis of patient's symptoms; an engineer, who detects the possibility that a definite block in a complicated technical system is the cause of malfunction on the basis of a set of indicators' data; an editor, who decides on the manuscripts according to the referees' evaluations, and others.

Usually one does not need to choose the best variant in them: the task is to categorize each object. So the final goal in such tasks is to distribute alternatives among classes of decision: to include or not this R&D project into a program; to accept, to correct or to reject the manuscript submitted to a scientific journal; to define the appropriate kind of the disease for a patient, and so on.

In spite of the fact that classification tasks are wide spread in human decision making, their theoretical investigation within the framework of the multicriteria decision making is scarce.De Montgolfier and Bertier (1978) described a procedure of generalizing sets of attribute values into ordinal categories of a more general criterion upon the decision maker's judgment. This task may be viewed as that of ordinal classification. ELECTRE TRI (Yu,1992) and ROBOT Technique (Bana e Costa ,1992) are the recent examples of an approach to ordinal classification based on the idea to assign multiattribute alternatives to ordered classes if it is found out that all the alternatives' values (separately) belong to the appropriate class.

Our approach to the solution of such tasks is based on the method of ordinal classification, named by the authors as ORCLASS (see Larichev et al., 1989; Larichev & Moshkovich, 1986, 1987, 1990, 1992, 1994).

The task formulation

The task of ordinal classification discussed in this chapter may be presented in the following way.

A decision maker has a final set of L decision classes and must assign to them a set of cases (or objects). These classes are ordered for a decision maker (DM) in the sense that each object placed in the first class is preferable to all objects placed in the second class, and so on. Each object can be characterized by values on each of Q criteria. Values upon criterion scales are presented to the decision maker in a verbal form. The decision maker rank orders each criterion scale from the most to the least preferable one. An example of criteria and classes for the case of evaluation of articles in a scientific journal is presented below.

CRITERIA FOR EVALUATING ARTICLES, SUBMITTED TO THE JOURNAL

Criterion A. Correspondence to the journal's outline
A1.The article is directly related to the journal's outline.
A2.The article has correspondence to the journal's outline.
A3.The article has rather low correspondence to the journal's outline.

Criterion B. Theoretical value of the results
B1.The results are of sound theoretical value in the field.
B2.The results have some theoretical value in the field.

B3. The results have no theoretical value in the field.

Criterion C. Practical value of the results
C1.The results are of high practical value.
C2.The results have practical value.
C3.The results have no practical value.

Criterion D. Errors
D1.There are no errors and inconsistencies in the article.
D2.There are some errors and inconsistencies in the article.
D3.There are many errors and inconsistencies in the article.

Criterion E. Quality of the text
E1.The article is written in a good language.
E2.The article is not nicely written.
E3.The article is awfully written.

DECISION CLASSES
Class I.   The article is to be published.
Class II.   The article is to be revised by the authors.
Class III.   The article is to be rejected.

As there are Q criteria, and each criterion has a given number of discrete values, we are able to form the set of all possible combinations of values in criteria space (Cartesian product of criterion values). A complete classification system is developed when an a priori construction of classification of all possible criteria space combinations is completed. When an experienced decision maker and a real decision context is used, this classification reflects the decision maker's rules and can be used for categorization of real alternatives (objects). For example, the editorial board of a scientific journal can construct the complete classification in the formed criteria space, and use the result to make a decision for each reviewed article.

The task can be represented formally in the following way:

Given:
1. $K = \{q_i\}$ $i=1,2,...,Q$ is a set of criteria;

2. n_q is the number of possible values on the scale of the q-th criterion ($q \in K$);

3. $X_q = \{x_{iq}\}$ is a set of values for the q-th criterion (the scale of the q-th criterion); $|X_q| = n_q$ $(q \in K)$;

4. $Y=X_1 * X_2 * ...* X_Q$ is a set of vectors $y_i \in Y$ of the following type

$$y_i=(y_{i1},y_{i2},...,y_{iQ}); \text{ where } y_{iq} \in X_q \text{ and } N = |Y| = \prod_{q=1}^{Q} n_q \text{ ;}$$

5. L - number of ordered decision classes.

Required:
on the basis of a decision maker's preferences (judgments) to build a reflection

F: $Y \Rightarrow \{Y_1\}$, $l=1,2,...,L$ such that $Y = \bigcup_{l=1}^{N} Y_l$; $Y_l \cap Y_k = \varnothing$ if $k \neq l$ (where Y_1 is a subset of vectors from Y, assigned to the l-th class).

As it was stated in previous chapters, one of the first tasks facing the decision maker is the rank-ordering of possible values for one criterion from the set K. As a result ordinal scales for criteria are formed, in which the first value x_{q1} upon the criterion q $(q \in K)$ is more preferable for a decision maker than the second value x_{q2} upon the same criterion and so on.

If we use natural numbers to denote values in the ordinal scale X_q for the q-th criterion, we shall obtain a modified scale $B_q = \{1,2,...,n_q\}$, where $b_{iq} < b_{jq}$, if x_{iq} is more preferable for the decision than x_{jq}. Thus, for each ordinal scale X_q we form the unique ordinal scale B_q, reflecting the decision maker's preferences for values from X_q.

This information from a decision maker defines an anti reflexive and transitive binary relation of strict preference (or dominance) P^0 on the set Y:

$$P^0=\{ (y_i,y_j) \in Y \times Y | \forall q \in K \ b_{iq} \leq b_{jq} \text{ and } \exists q^0 \text{ such that } b_{iq}^0 < b_{jq}^0 \}.$$

Table 5.1 presents four hypothetical variants of articles submitted to the scientific journal, estimated against the criteria given above.

These alternatives may be presented through the following b-vectors according to the above described principle:

vector $b_1=(1,2,1,1,1)$ for alternative #1,
vector $b_2=(2,2,1,2,1)$ for alternative #2,

vector b_3=(2,2,3,2,1) for alternative #3,
vector b_4=(2,1,1,1,1) for alternative #4.

Table 5.1. Data for four hypothetical alternatives

Alternatives (articles)	Correspondence to the journal's outline	Theoretical value	Practical value	Errors	Quality of the text
#1	direct	some	high	no	good
#2	relative	some	high	some	good
#3	relative	some	no	some	good
#4	relative	sound	high	no	good

As we can see alternative #1 dominates alternative #2 (as it has better values upon criteria one and four), alternative #2 dominates alternative #3 (as it has better value upon criterion three), and alternative #4 dominates alternatives #2 and #3. Alternatives #1 and #4 are incomparable upon dominance relation.

On the other hand we know that decision classes are ordered for the decision maker. This means that any alternative from the first class is more preferable for the decision maker than any alternative from the second class and so on. This property may be reflected in the following binary preference relation on the set Y:

$$P^1=\{(y_i, y_j) \in Y \times Y | \ y_i \in Y_k, y_j \in Y_l, k < l\}.$$

It is natural to assume that no vector from Y dominating the one under consideration, is to be in a less preferable class. For our example this means that if the alternative #2 from Table 5.1 belongs to the first class according to the decision maker's opinion, then alternative #1 must also belong to the first class, as it dominates the alternative #2. Formally this may be put down as follows:

if $(y_i, y_j) \in P^0$ and $y_i \in Y_l$, then $y_j \notin Y_k$ if $k < l$.

Let call the partition of the set Y into classes non-contradictory if this requirement is fulfilled. The requirement for the partition to be non-contradictory is equal then to the fulfillment of the following:

$$\text{if } (y_i, y_j) \in P^0, \text{ then } (y_j, y_i) \notin P^1$$

$$(5.1).$$

Definition 6.1. The partition of the set Y into L rank ordered classes is called *non-contradictory* if for any y_i, $y_j \in Y$ the condition 5.1 is fulfilled.

As a result, it is necessary to construct a complete and non-contradictory classification into L rank ordered classes of all objects from Y on the basis of the decision maker's preferences.

An approach to the rational procedure for construction of the complete ordinal classification

It is possible to accomplish the task of classification by having the decision maker classify directly all possible vectors from Y. However, this is impractical even for a relatively small problem, which can involve a large number of such vectors. Therefore, a special procedure for elicitation of decision maker classification rules has been developed (Moshkovich, 1985).

The proposed procedure allows to construct the needed classification with the help of a limited number of questions to a decision maker. The idea of the procedure is based on the implementation of the requirement (5.1).

Suppose we have only two decision classes for our example: class I means that the article is to be published, and class II means that the article is to be rejected. If we ask the decision maker to classify alternative #2 we shall be able to classify alternatives #1 or #3 without presenting it to the decision maker. If the decision maker considers alternative #2 worth of class I, then alternative #1 is also to be assigned to class I (as it dominates alternative #2). In case, the decision maker considers that alternative #2 is to be rejected (class II), we are able to say that alternative #3 is also to be rejected (as it is dominated by alternative #2). Thus, any answer of the decision maker for the alternative #2 will determine the appropriate class for one of the other two alternatives from our example.

Note, that if first we present the alternative #1 or alternative #3 for classification the consequences may be different. In case, we present alternative #1, and the decision maker assigns it to class I (and this is very probable as it has good attribute values) we shall not be able to make any conclusions about appropriate the class for alternative #2 or alternative #3. Analogous result will be obtained if we present alternative #3 and the decision maker assigns class II to it.

Therefore, it is attractive to classify as many alternatives as possible by logical rules inferred from previous classifications given by the decision maker. Thus, the choice of a vector from Y for classification by the decision maker may influence the effectiveness of the interview (if effectiveness is evaluated by the number of vectors from Y the decision maker has to directly classify to complete the whole task). In this sense vectors from Y may be differently "informative" for the construction of a complete classification of vectors from Y, and we can formulate a task of determining the most "informative" vector at each step of the dialogue with the decision maker.

Let denote G_i the set of class numbers, reflecting classes Y_l ($l=1,2,...,L$), permissible for vector $y_i \in Y$. Before the dialogue with the decision maker starts, for each $y_i \in Y$ $G_i = \{1,2,...,L\}$, as we have no relevant information about the decision maker's preferences. As the goal of the dialogue with the decision maker is to define a single class for each vector from Y, it is necessary in the end to have only one element in each G_i, $|G_i| = 1$ for each $y_i \in Y$.

Let the decision maker to assign vector y_i to class Y_l ($1 \leq l \leq L$), this is denoted as $y_i \in Y_l$. It is clear, that in this case the vector with criterion values not less attractive than in y_i must not belong to a less preferable class (according to 5.1). This can be put down as:

$$\text{if } y_i \in Y_l \text{ and } (y_j,y_i) \in P^0, \text{ then } y_j \notin Y_k, \text{ where } k > l$$

(5.2).

Analogously, vector with criterion values not more attractive than in y_i, must not belong to a more preferable class. This may be put down as:

$$\text{if } y_i \in Y_l \text{ and } (y_i,y_j) \in P^0, \text{ then } y_j \notin Y_k, \text{ where } k < l$$

(5.3).

These two expressions show how information about the appropriate class for one vector from Y may be used to restrict the set of permissible classes (and sets G_j correspondingly) for other vectors from Y.

The number of vectors presented to the decision maker for classification (to construct complete classification of vectors from Y) may be decreased through implementation of binary relations P^0 and P^1, defined on the set Y. The number of indirectly classified vectors depends on the vector presented to the decision maker for classification, and on the class assigned by the decision maker to the presented vector. To evaluate the possible amount of information obtained through classification by the decision maker of a vector from Y, it is necessary to calculate the number of indirectly classified vectors for each possible class for the presented vector.

Let g_{il} denote the number of vectors definitely classified by assigning class Y_1 to vector y_i. Thus, g_{il} characterizes the amount of information gained as a result of such decision. This amount depends on the class prescribed to the vector y_i. As we do not know in advance the class the decision maker will assign to the presented vector, it is reasonable to attempt while evaluating the possible amount of information connected with the vector y_i, to introduce some index which will characterize the likelihood of class Y_1 for vector y_i . We propose the following heuristic approach to this problem.

Let p_{il} denote the index which reflects the likeliness of y_i, being assigned class Y_1. Then the expected amount of information Φ connected with classification of vector y_i may be defined as:

$$\Phi_i = \sum_{l \in G_i} p_{il} g_{il} \tag{5.4}$$

The presented ideas formed the basis for the procedure, proposed by Larichev et al.,(1980), Moshkovich (1985), which minimized the number of vectors presented to the decision maker for classification for the case L=2. For case L > 2 a stepwise procedure was proposed (Moshkovich, 1985; Larichev & Moshkovich, 1986). The procedure guarantees the maximum of expected information at each step the dialogue with the decision maker. Let analyze it in a more detail.

There may be different heuristics for calculation of p_{il} in the formula (5.4). It is clear that the possibility of vector y_i to be assigned class Y_1 is connected with some notion of the "similarity" of y_i and elements of class Y_1. We introduce the formal idea of the center c_1 of class Y_1, which is defined according to formula (5.5). This is an artificial point in the criteria space with averaged values upon all criteria. Though it has no special physical sense it reflects some averaged image of the class representative, and will be used later to evaluate the required "similarity":

$$c_l = (c_{l1}, c_{l2}, \ldots, c_{lQ}), \text{ where } c_{lq} = \sum_{y_i \in Y_l} y_{iq} / |Y_l|, \; q=1,2,\ldots,Q \qquad (5.5)$$

<u>Definition 5.2.</u> Let us call the *center of a non-empty class* Y_l the point in a multidimensional space $c_l = (c_{l1}, c_{l2}, \ldots, c_{lQ})$, each component of which is equal the average of corresponding components of vectors from Y belonging to class Y_l (see 5.5).

Let require for the coordinates of the center for the empty class Y_l to be different from the corresponding coordinates of the centers of classes Y_{l-1}, and Y_{l+1} on one and the same value. Let $Y_s \neq \emptyset$, $Y_t \neq \emptyset$, and for each l such that $s<l<t$ $Y_l = \emptyset$. Then coordinates for centers of empty classes will be defined according to the following formula:

$$c_{lj} = c_{sj} + \eta_j(1-s), \text{ where } \eta_j = (c_{tj} - c_{sj})/(t-s).$$

The index p_{il} then, is based on the measure of the "distance" between the vector y_i and the center c_l. Smaller distance will reflect larger possibility for vector y_i to be assigned class Y_l. Therefore, distance d_{il} between vector y_i and the center of the l-th class will be calculated upon the following formula:

$$d_{il} = \sum_{q=1}^{Q} |y_{iq} - c_{lq}|.$$

Let d_{max} denote the maximum possible distance between two vectors from Y:

$$d_{max} = \sum_{q=1}^{Q} (n_q - 1),$$

where n_q ($q=1,2,\ldots,Q$) is the number of criterion values on the scale of the q-th criterion.

<u>Definition 5.3.</u> Let call p_{il} the *similarity measure* for vector $y_i \in Y$ and class Y_l if it is calculated as:

$$p_{il} = (d_{max} - d_{il}) / (\left|G_i\right| d_{max} - \sum_{s \in G_i} d_{is}).$$

The formula shows that $0 \le p_{il} \le 1$ and $\sum_{l \in G_i} p_{il} = 1.$. This index becomes larger as the distance between the vector y_i and class Y_l becomes smaller (it is considered that in this case the probability that vector y_i will be assigned class Y_l is larger).

As we have ordinal criteria scales, the introduced formulas may not be quantitatively meaningful, but as their outputs are used only in a substantive sense (to generate some rough estimation of our expectations), we consider them enough valid and useful in the proposed heuristic procedure.

The rational procedure of interviewing a decision maker is based on sequential presenting to the decision maker the most informative vectors from Y, that is y_i for which:

$$\Phi_i = \max_{y_i \in Y_0} \Phi_j \quad Y_0 = \{y_j \in Y \mid \left|G_j\right| > 1\}.$$

Let y_i be the most informative vector at the current step. We present it to a decision maker for classification, and receive the answer that $y_i \in Y_l$. It is natural to assume that all $y_j \in Y_0$ such that $(y_j, y_i) \in P^0$, y_j may not belong to the class less preferable than Y_l. This will require the corresponding modification of G_j. Analogously for each $y_j \in Y_0$ such that $(y_i, y_j) \in P^0$, y_j may be assigned class not more preferable than Y_l. This will also require corresponding modification of G_j.

Thus, the rational procedure for the dialogue with the decision maker may be presented in the following general steps:

1. The subset $Y_0 \subset Y$ is defined: $Y_0 = \{y_j \in Y \mid |G_j| > 1\}$. If $Y_0 = \varnothing$, go to step 7.
2. For each $y_j \in Y_0$ the index p_{il} is calculated according to the definition 5.3 for each $l \in G_j$.
3. For each $y_j \in Y_0$ the index Φ_j is calculated according to formula 5.4.

4. $y_i \in Y_0$ is defined such that: $\Phi_i = \max\limits_{y_j \in Y_0} \Phi_j$

5. Vector y_i is presented to the decision maker for classification.

6. In accordance with the class Y_1 assigned by the decision maker to the vector y_i subsets G_j are modified in the following way:

    for each $y_j \in Y_0$ such that $(y_i,y_j) \in P^0$ let define G_j as: $G_j \cap \{1,2,...,l\}$;
    for each $y_j \in Y_0$ such that $(y_j,y_i) \in P^0$ let define G_j as: $G_j \cap \{l,l+1,...,Q\}$.
    Go to step 1.

7. The procedure is over.

The proposed procedure uses heuristic approach and therefore needs some evaluation. To evaluate the effectiveness of the procedure, statistical modeling was used.

To generate the initial partition of the set Y on L classes the following algorithm was proposed and used. The researcher defines the number of criteria Q, the number of criterion values on a criterion scale n_q, and number of classes L. The Cartesian product of scales (the set Y) is formed. It is known that the vector with all the first (most preferable) criterion values belongs to class Y_1, and the vector with all the last criterion values (least preferable ones) belongs to class Y_L. Then according to the described procedure, the vector y_i is defined (with the maximum expected amount of information if classified). The decision maker's response is simulated with the help of the random numbers' generator: indices p_{il} calculated according to the definition 5.3 are used as probabilities for the vector to be assigned corresponding class. According to the number r obtained through the random number generator, class for vector y_i is determined as:

$$y_l \in Y_l, \text{ if } \sum_{t=0}^{l-1} p_{it} < r < \sum_{t=1}^{l} p_{it}, \text{ where } p_{i0} = 0.$$

In accordance with the dominance relation P^0 subsets G_j are modified for each vector y_j from the set Y, which class has not been defined yet. The centers of all classes are recalculated as well as indices Φ_j for non-classified vectors from Y. The vector y_i with the maximum Φ_i is selected for classification.

According to the random number appropriate class for it is defined, and so on. The procedure is carried out till all vectors from Y are classified. The number of used random numbers (for simulating decision maker's responses) gives us the number of classifications (N_0) needed from the decision maker to construct the complete classification for the task using the proposed rational procedure.

For each variant of Q, n, and L about 1000 simulations were carried out, and the number N_0 were registered.

The averages values for this data for cases with four and five criteria with three and four criterion values on the scales, and two, three, or four classes are given in Table 5.2.

Table 5.2. Averages for the number of classifications (N_0) required from the decision maker using the rational procedure for different task dimensions

Number of criteria (Q)	Number of criterion scales (n_q)	Size of the Y (N)	Number	of classes	(L)
			2	3	4
4	3	81	8	13	17
4	4	256	10	14	21
5	3	243	10	18	25
5	4	1024	14	24	33

The data shows that the proposed procedure decreases the number of classification required from the decision maker essentially comparing with the size of the set Y. This supports the idea that this procedure is rather effective for interviewing a decision maker in an ordinal classification task.

Detection and elimination of inconsistencies in the decision maker's responses

In accordance with the requirements formulated in chapter 2 of this book, the decision making method is to provide verification of the decision maker's judgments for consistency.

In our case, the possibility for detection of errors is based on the rank ordering of criterion values and decision classes, expressed in relations (5.2) and (5.3). Violations of these requirements indicate the presence of some inconsistencies in the gained responses, as this relation requires more preferable objects to be assigned to a more preferable class. For example, if alternative #1 (see Table 5.1) is assigned to the second class, and alternative #2 has been assigned to the first class, we can conclude that these two responses contradict each other, as alternative #1 has better criterion values. The decision maker must reconsider information and change one of the (or both) responses.

There may be two main reasons for inconsistencies in received responses:

1. there are errors in rank ordering of criterion values (upon criterion scales);

2. there are errors in the decision maker's responses concerning classification of presented alternatives.

Let consider that the dominance relation is true, and, as a result, the inconsistencies are caused by incorrect responses of the decision maker (while classifying some of the presented vectors).

Definition 5.4. Let call an *error in the built classification* the violation of the requirement (5.2) or (5.3), which means the situation when a less preferable class is assigned to a more preferable vector.

In this case we shall say that the decision maker's responses contradict to each other. We suggest the following approach to the elimination of inconsistencies.

Let $Y^0 \subset Y$ be the subset of vectors being classified up to the moment. Now, the decision maker is presented with vector $y_i \in Y$. G_i is the set of possible classes for y_i according to the information at hand, that is $G = \{l, l+1, ..., l+m\}$, where $l \geq 1$ and $l+m \leq L$.

To get better understanding of the process consider data in Table 5.1. If we have three decision classes, and the decision maker has assigned the second class to alternative #2, we are able to conclude that alternative #1 may belong to the second or to the first class (as it dominates alternative #1): $G_1 = \{1, 2\}$. At the same time alternative #3 may belong to the second or the third class, and, as so, $G_3 = \{2, 3\}$ (as alternative #2 dominates alternative #3).

Let us suppose that the decision maker assigns class Y_s to y_i, and $s < 1$ or $s > 1+m$ (e.g. alternative #1 in our example is assigned the third class, or alternative #3 is assigned the first class). In this case the response *is inconsistent* with the previous one(s), as there exists at least one vector in Y^0 dominating y_i, and assigned to class Y_1 (that is why possible classes for y_i in G_i start from 1). Analogously, there exists a vector in Y^0, which is dominated by y_i and is assigned class Y_{1+m}.

Thus, if the decision maker's response is not within the subset G_i, it means one of the following:

1. if $s < 1$, then there exists $y_j \in Y^0$ such that $(y_j, y_i) \in P^0$ and $y_j \in Y_z$, $z > s$;

2. if $s > 1+m$, then there exists $y_j \in Y^0$ such that $(y_i, y_j) \in P^0$ and $y_j \in Y_z$, $z < s$.

Previous considerations may be used for construction of a subset of classified vectors from Y, violating relation (5.2) or (5.3). Let us denote such subset as Y_{err}, and define it's elements in the following way:

$$s < 1: Y_{err} = \{y_j \in Y^0 \mid (y_j, y_i) \in P^0, y_j \in Y_z, y_i \in Y_s, z > s\} \qquad (5.6)$$

$$s > 1+m: Y_{err} = \{y_j \in Y^0 \mid (y_i, y_j) \in P^0, y_j \in Y_z, y_i \in Y_s, z < s\} \qquad (5.7)$$
$$(6.7)$$

For our example Y_{err} will contain only one vector, that is alternative #2. In general case, this set may be large enough.

Then the decision maker is consequently presented with pairs of vectors, (y_i, y_j), where $y_j \in Y_{err}$, and classes assigned to them. The decision maker is to analyze the contradiction and to change one or both of these responses to eliminate it. After that, the correction of the appropriate classes in accordance with new assignments is carried out, and y_j is eliminated from Y_{err}. When $Y_{err} = \varnothing$, the procedure stops.

It is necessary to note that when we have more than two classes of decision, the newly given responses may be different. That is why the elimination of all elements from Y_{err} does not guarantee the absence of contradictions in the whole classification of the set Y^0 (that is: new classes for some vectors may be inconsistent with classes of other vectors from Y^0, previously assigned to them).

Let us illustrate it by the example for data from Table 5.1. Let us consider that we know classes of alternatives #2, #3, and #4. Alternatives #2 and #4 belong to the second class, and alternative #3 is assigned the third class. This gives us $G_1=\{1,2\}$ for alternative #1. The decision maker marks the third class as appropriate for this alternative. A contradiction appears, and $Y_{err}=\{$alternative #2$\}$. We present the decision maker with alternatives #1 and #2, and request the necessary change of the received responses. The decision maker analyzes the situation, and decides that alternative #1 and alternative #2 must both belong to the first class. In this case, we eliminate the previous contradiction ($Y_{err} = \varnothing$), but there appears a new one: alternative #4 dominates alternative #2 and belongs to the second class, while alternative #2 is now assigned to the first class.

Thus, only if the decision maker changes the class for y_i, we are able to continue our procedure. Otherwise, it is necessary to check requirements (5.6) and (5.7) for all $y_j \in Y^0$ with changed classes. If new contradictions appear, the procedure is to be repeated.

To make the task less complex we suggest in this case to build a matrix of contradictions A. Let I be a capacity of the set Y^0: $I=|Y^0|$. Let rearrange vectors in Y^0 in a way for the following vectors not to be more preferable (according to dominance relation) than the preceeding ones. This can be easily done by presenting vectors in a lexicographic order (an ascending order of criterion values starting with the last criterion).

Then $A=\| a_{ij} \|$ of the dimension I x I in which:

$$a_{ij} = \begin{cases} 1, \text{ if } (y_i, y_j) \in P^0, i, j = 1,2,...,I, \text{ and } i \neq j, \\ \hline 0, \quad \text{otherwise} \end{cases}$$

As all vectors are arranged in an ascending lexicographic order, all a_{ij}, equal to 1, will be placed over main diagonal (so , it is possible to present only this part of the matrix). Such matrix for our example is presented as matrix AI.

Let us underline a_{ij} equal to 1, for which $y_i \in Y_s$, $y_j \in Y_z$ and s > z (matrix with underlined elements for our example is matrix AII).

(AI)

Vector from Y⁰		#1	#4	#2	#3
	Class number	III	II	II	III
#1	III		0	1	1
#4	II			1	1
#2	II				1
#3	III				

(AII)

Vector from Y⁰		#1	#4	#2	#3
	Class number	III	II	II	III
#1	III		0	<u>1</u>	1
#4	II			<u>1</u>	1
#2	II				1
#3	III				

It is evident that these underlined elements reflect contradictions in the decision maker's responses. If there are no marked elements, we are able to continue the interview to build the classification. Otherwise, it is necessary to present the decision maker with a pair of vectors corresponding to the underlined element, correct the class, and correct the corresponding information in the matrix A. You can see the modified matrix A after assigning the first class to alternatives #1 and #2 (matrix AIII).

(AIII)

Vector from Y⁰		#1	#4	#2	#3
	Class number	I	II	I	III
#1	I		0	1	1
#4	II			<u>1</u>	1
#2	I				1
#3	III				

Thus the correction process is fulfilled, only if there are no underlined elements in the matrix A. Otherwise, the underlined elements are being processed in the same manner.

Let analyze in what cases new contradictions can appear. Let $y_i \in Y_s$, and $y_j \in Y_{err}$ (where Y_{err} is formed according to (5.6) and $y_j \in Y_z$). When presented to the decision maker for the analysis, the vector y_j was reclassified as belonging to class Y_t, where $t \le s < z$ (to eliminate the contradiction). New contradiction can appear only if there exists $y_k \in Y^0$ such, that $(y_k, y_j) \in P^0$ and $y_k \in Y_l$, where $l > t$ (but $l \le s$, as otherwise, y_k would belong to Y_{err} and the number of the appropriate class for y_k would be decreased in the process of elimination of inconsistencies).

Analogously, $y_j \in Y_{err}$ (where Y_{err} is formed according to (5.7) and $y_j \in Y_z$), the new contradiction can appear only if the vector y_j is reclassified as belonging to class Y_t, where $t \ge s > z$ there exists $y_k \in Y^0$ such, that $(y_j, y_k) \in P^0$ and $y_k \in Y_l$, where $l < t$ (but $l \ge s$, as otherwise, y_k would belong to Y_{err} and the number of the appropriate class for y_k would be increased in the process of elimination of inconsistencies).

The convergence of such sequential correction procedure is guaranteed by the fact that at each iteration the decision maker is to decrease the class number for vectors from Y_{err} (in case $s < l$, case (5.6)), or consequently to increase it (in case $s > l+m$, case (5.7)). In these circumstances the finite, and not large number of possible classes limits the possibility of appearance of new contradictions, providing a high speed of convergence.

After the elimination of all contradictions in decision maker's responses, it is necessary to simulate the procedure of processing these responses for vectors from Y, as if obtained newly from a decision maker. This will allow to correct sets G_j for all non-classified vectors from Y.

Thus, the suggested approach allows to carry out an effective procedure of "on-line" correction of possible errors in the decision maker's responses while constructing complete ordinal classification of vectors from the Y for any number of decision classes.

Class boundaries and their implementation

Proposed procedures for selection of informative questions to the decision maker, and detection and elimination of inconsistencies in the decision maker's responses, allow to organize an effective procedure for construction of the complete and contradiction free ordinal classification of vectors from Y on the basis of the decision maker's preferences. As a result we obtain a partition of the set Y among L classes, satisfying the initial requirement of the task under consideration. This partition may be used to determine the appropriate decision class for any object described by the formed set of criteria.

At the same time, the constructed classification may be used for illustration of decision rules, used by the decision maker while classifying vectors from Y.

Let analyze in a more detail elements of one class Y_1. It is clear (according to the process of the construction of the classification), that any two elements of the same class y_i and y_j may be related to one another according to dominance relation, or be incomparable. This means that one of the following three expressions is to be true:

1. $(y_i, y_j) \in P^0$;

2. $(y_j, y_i) \in P^0$;

3. $(y_i, y_j) \notin P^0$; and $(y_j, y_i) \notin P^0$.

The last condition means that vectors y_i and y_j are incomparable with one another and will be denoted as $y_i \sim y_j$.

Thus, in class Y_1 it is possible to elaborate subsets of non-dominated and dominated elements $Y_1^{\,1}$ and $Y_1^{\,2}$ correspondingly:

$Y_1^{\,1} = \{y_i \in Y_1 |$ for any $y_j \in Y_1 \Rightarrow (y_i, y_j) \in P^0$ or $y_i \sim y_j\}$

$Y_1^{\,2} = \{y_i \in Y_1 |$ for any $y_j \in Y_1 \Rightarrow (y_j, y_i) \in P^0$ or $y_i \sim y_j\}$.

Let call these subsets as *class boundaries* for class Y_1.

Definition 5.5. The set of all non-dominated elements from class Y_1 (l=1,2,...,L) will be called the *upper boundary of the class*. The set of all non-dominating elements of class Y_1 will be called the *lower boundary of the class*.

Thus, according to the definition 5.5, the set $Y_1^{\,1}$ is the upper boundary of class Y_1, and the set $Y_1^{\,2}$ is the lower boundary of the class Y_1. What does this definition mean? It means that all vectors from Y, which are not more preferable than elements of the upper boundary of the class Y_1, and which are at the same time, not less preferable than elements of the lower boundary of the class Y_1 belong to class Y_1. In this sense, class boundaries define the rules according to which vectors from the set Y are assigned appropriate classes. Let illustrate this using an example.

Let consider that in our example of articles' evaluation the decision maker carried out all the necessary classifications, and the compete and contradiction free classification of all possible articles is built, using four decision classes (the second class for the articles to be revised was divided into two):

Class I: the article is to be published;
Class II: the article is to be revised and then published;
Class III: the article is to be revised and then reviewed once again;
Class IV: the article is to be rejected.

On the basis of the built classification upper and lower boundaries for all four classes were elaborated in accordance with the definition 5.5. In figure 5.4 elements of these boundaries are given in the form of b-vectors (to be more understandable).

Let try to find the appropriate class for each of our four alternatives, using these boundaries. Note, that alternative #1 is characterized by b-vector (12111), alternative #2 is described by b-vector (22121), alternative #3 has b-vector (22321), and alternative #4 has b-vector (21111).

First, let analyze alternative #1. It dominates elements of the lower boundary of class I, and thus, is to be assigned class I. Alternative #2 has second value against the fourth criterion (ERRORS), and thus does not belong to class I. As it dominates the elements of the lower boundary of the class II, it is to be assigned class II. Alternative #3 does not belong to class I (as it also has second value against the fourth criterion), but belongs to class II as it dominates the element of the lower boundary of this class. Alternative #4 belong to class I as it dominates elements of the lower boundary of this class.

This example shows that boundary elements are enough to be able to assign appropriate class for any possible vector from the set Y. This process can be easily understood and implemented. The decision rules used for classification can be illustrated through these boundary elements, providing efficient and understandable way of presenting classification rules, used by the decision maker. The efficient set of production rules corresponding to the presented boundary elements of the example (see figure 5.4) is presented in figure 5.5.

Figure 5.4. Visualization of class boundaries (the most and least preferable vectors in each class)

CLASS I
Elements of the upper boundary of the class:
11111
Elements in the lower boundary of the class:
23212   23311

CLASS II
Elements of the upper boundary of the class:
11121   11113   11312
Elements of the lower boundary of the class:
23323

CLASS III
Elements of the upper boundary of the class:
11131
Elements of the lower boundary of the class:
13333   22333   23233   23332

CLASS IV
Elements of the upper boundary of the class:
31111   23333
Elements of the upper boundary of the class:
33333

Presentation of the classification rules as sets of boundary elements thus allows to formulate efficient set of decision rules as well as to carry out quick analysis and modification of classification rules in real situations.

Figure 5.5. Classification rules for the articles' evaluation

---

RULE 1:
If the article has "low correspondence to the journal's outline"
THEN article is "TO BE REJECTED"

RULE 2:
If the article has "some correspondence to the journal outline" AND has "no theoretical value" AND has "no practical value" AND has "has essential errors" AND has been "awfully written"
THEN article is "TO BE REJECTED"

RULE 3:
If the article is NOT "TO BE REJECTED" AND has "essential errors"
THEN the article is "TO BE REVISED AND THEN REVIEWED ONCE AGAIN"

RULE 4:
If the article is NOT "TO BE REJECTED" AND the article is NOT "TO BE REVISED AND THEN REVIEWED ONCE AGAIN" AND has "some errors and inconsistencies" OR "awfully written"
THEN the article is "TO BE REVISED AND THEN PUBLISHED"

RULE 5:
If the article is "NOT TO BE REJECTED" AND the article is NOT "TO BE REVISED AND THEN REVIEWED ONCE AGAIN" AND has "no practical value" AND has been "written not nicely"
THEN the article is "TO BE REVISED AND THEN PUBLISHED"

RULE 6:
If the article is NOT "TO BE REJECTED" AND the article is NOT "TO BE REVISED AND THEN REVIEWED ONCE AGAIN" AND the article is NOT "TO BE REVISED AND THEN PUBLISHED"
THEN the article is "TO BE PUBLISHED"

Behavioral aspects of the ordinal classification task

As was discussed in chapter 3, human capabilities in different judgmental operations are limited. The proposed method ORCLASS is based on the direct classification by a decision maker of complicated multicriteria objects (alternatives). This may cause rather large load on a human short-term memory. That is why a series of investigations of the human capacities in ordinal classification task were carried out (Larichev & Moshkovich, 1980, 1988, 1990, 1992, 1994; Larichev et al., 1980, 1988).

General schema of experiments

Let discuss results of two series of experiments analyzing the behavior of three groups of subjects: senior college students, senior high school students, and members of the editorial board of a journal in a large research institute (Larichev et al., 1988).

The complexity of the classification problem in each experiment was determined by three parameters: the number of criteria, characterizing the evaluated objects (Q); number of values on their scales (n_q), rank ordered from the best ot the worst; and number of ordered decision classes (L), to which the considered objects should be assigned. The hypothesis was that human behavior varies as a function of changes in these problem variables.

In the course of the experiments, subjects were requested to classify all possible objects (vectors from Y), using the prescribed decision classes.

The subjects' behavior was evaluated against three criteria:

1. Number of inconsistencies (errors). It was viewed as violation of the requirement (5.1) discussed in this chapter (less attractive alternative is assigned to a more attractive decision class).

2. Number of substitutions. Along with the number of inconsistencies, information is also provided on the changes that must be made in the subject's answers in order to make the classification consistent. It is common knowledge that people often make errors in information processing operations. No two errors are alike, however. Errors made far from class boundaries (see figure 5.4) are evident (they lead to a large number of inconsistencies and may be easily found as the reason

for inconsistencies without changing boundaries between classes). Errors, made in the vicinity of class boundaries are quite another matter. In this situation only two vectors contradict each other, it is difficult to decide what classification is not correct, and each decision leads to modifications in class boundaries. Thus, the number of subtitutions was calculated for each subject.This number could be much less then the number of inconsistencies in the cases of several big errors.

3.   Complexity of the decision rule. This characteristic (Larichev & Moshkovich, 1980, 1988) implies the complexity of the rules the subject employed in classification. If class boundaries contain rather large number of elements, which are described by a large number of complicated production rules, then it is considered that the subject was able to use complicated decision rules to construct complete classification of vectors from the set Y.

It is necessary to distinguish between the experiments involving people with almost no decision making experience (students in the first series of experiments) and those involving professionals handling real-life problems (second series of experiments).

For the first group of subjects there were ample opportunities for modifying the classification problem parameters and terms of experiment. College students (experiments from 1 to 12) classified possible apartments for rent, deciding to what extent the presented alternatives satisfied them. High school students (experiments 13 and 14) classified universities with the regard for their suitability for entering after graduation from the high school.

For the second group of people, when classification was a real decision making problem, the opportunities for modifying the problem parameters were almost completely deprived because the experimental pattern corresponded to a real problem. The second series of experiments involved members of an editorial board, who estimated the quality of manuscripts submitted for publication (experiments 15 and 16).

Results of the first series of experiments

Data on the average number of errors made by the subjects during classification of 100 different vectors from the set Y in each of the 14 experiments of the first series are shown in table 5.4. As is seen from the table, the average number largely depends on the complexity of the classification problem. An ANOVA on the number of errors as a function of parameters Q and n_q showed that under constant Q : n_q ratio the number of errors (inconsistencies) depends on the number of decision classes L.

Table 5.4. The parameters of classification problem and major results in various sets of experiments

Experiment	Number of subjects	Q	n_q	L	N	E	%
First series							
1	9	7	2	5	128	9.5	11
2	9	7	2	4	128	6.5	0
3*	19	7	2	3	128	6.5	37
4	15	5	3	4	243	9.7	13
5*	20	5	3	3	243	5.8	35
6*	24	5	3	2	243	5.0	46
7	20	4	4	4	256	8.8	10
8	20	4	4	3	256	6.2	20
9*	9	4	4	2	256	3.0	67
10	10	3	5	5	125	17.0	0
11	10	3	5	5	125	8.8	9
12*	10	3	5	3	125	5.1	60
13	16	5	3	4	243	9.8	19
14*	16	5	3	2	243	3.5	73
Second series							
15*	9	5	3	2	243	3.3	-
16	4	5	3	4	243	1.3	-

Note: Q - number of criteria; n_q - number of criterion values on a scale; L - number of decision classes; N - number of vectors in the set Y (number of classified objects); E - average number of errors made on 100 classified vectors from the set Y; % - percentage of subjects managing the task; * - complexity of the problem is within human capabilities.

The average time for assigning an object for a class amounted to 14 sec. An additional analysis of classification quality, conducted for each subject against criteria "Number of errors","Number of substitutions" and "Complexity of the decision rule", made it possible to determine whether the subject managed the task properly.

The requirements were: to be considered as a subject who managed the work, it was necessary to have not more than two errors within the boundaries vicinity, and to have at least one complicated production rule used for classification (not to convert the problem into "elimination by aspects" viewing criteria as simple constraints).

The percentage of subjects who succeeded in accomplishing the task in each experiment is also shown in the table (%). Thus, for experiment 9, for example, for Q=4, n_q=4 (for all criteria) and L=2, 67% of subjects managed the job (the average number of errors was 3). For the same Q=4 and n_q=4 but for L=4, 90% of subjects *failed* to accomplish the task and there was a marked increase in the number of inconsistencies and errors (average number is 8.8).

It was also conventionally agreed that if at least one third of the group of subjects successfully managed the task, the classification problem with given parameters was within the human abilities. The experiments, for which in conformity with our criteria, the complexity of classification problem is within the subjects' capacities, are marked with an asterisk (*).

The experimental results confirmed the hypothesis that there are certain 'limits' to the subjects' capacities in multicriteria ordinal classification problems (Larichev et al., 1988). The available results indicate that for some parameter values the number of inconsistencies and errors sharply increase. The subjects fail in managing the problem, and their answers make it impossible to draw boundaries between classes.

Analysis and discussion of results of the second series of experiments

The major purpose of these experiments was to see how professionals managed classification problems and to what extent the results obtained with groups of students may be related to real-life classification problems.

Two experiments were conducted in which a multiattribute classification problem was combined with the occupational tasks of the subjects. In experiment 15 the subjects classified manuscripts submitted for publication ($Q=5$, $n_q=3$, $L=2$). The average number of errors is given in table 5.4. The analysis showed that all but one subject (members of the editorial board) managed the classification problem with two classes of decision. The analysis of class boundaries indicated that no subject made use of more than five rules during classification.

We expected that as the problem becomes more complicated (experiment 16: 4 decision classes), the number of errors should decrease. The number of errors turned out to decrease (see table 5.4). Of special interest in this connection is the analysis of subjects' strategies.

The analysis of class boundaries showed that as the problem became more complicated, the subjects simplified their strategies (decision rules). As a result, two out of four subjects, according to the "complexity of the decision rule" criterion, failed to solve the classification problem in spite of small number of errors made. It is worth noting that although the substantive content of the classification problem was familiar to the subjects, it was presented in a form of multicriteria description (not real manuscripts), that was unusual to them.

Thus, we may conclude that the behavior of experienced decision makers differs from that of the ordinary people when handling increasingly complicated problems. First, they try, primarily, to be consistent. In doing so, they simplify their task by discarding some criteria from consideration and transferring them to constraints. Having substantially simplified the original problem they in fact solve a different problem that is more amenable to human information processing capabilities.

The main reason for such behavior from our point of view, is the limited capacity of the human short-term memory. While solving recurring classification problems experienced decision makers use the already existing in the long-term memory structural information units. The number of these units used simultaneously is limited by the limits of the short-term memory. At the same time the structural units themselves may be rather large.

It is reasonable to assume that the capabilities of experienced decision makers have the same limitations. Understandable difficulties in carrying out experiments with these category of subjects caused the small number of experiments, which makes it difficult to make definite conclusions. Results of all experiments are summarized in table 5.5, where the marginal number of criteria under which subjects still manage multicriteria classification problems is shown (Larichev et al., 1986).

Table 5.5. The marginal number of criteria (Q) under which the subjects still manage solution of new multicrtieria classification problems

Number of criterion values on ordinal scales (n_q)	Number of decision classes(L)			
	2	3	4	5
2	7-8	6-7	4-5	3
3	5-6	3-4	2-3	2
4	2-3			

In other ranges of task parameters, number of inconsistencies sharply increased. Subjects failed in managing the problem, and their responses did not allow to elaborate meaningful decision rules of classification.

The analysis showed that when the subjects managed the task the number of rules they used did not exceed eight. In cases when they failed, a formal analysis revealed a much larger number of rules.

The most suitable explanation for the above data is probably as follows. In assigning an alternative to some class or other, the subject has to keep all rules in short-term memory, constituting structural units of information (chunks) he (or she) operates. As is known, the volume of short-term memory is limited. Different scientists (Miller 1956, Simon 1974, 1981) indicate that it does not exceed 5-9 structural units of information (chunks), and they may differ in size.

When subject employed 9 or fewer rules for classification, they managed the task. If more, then a part of rules were abundant for the operating in short-term memory which sharply increased the number of errors and inconsistencies.

The data presented in table 5.5 mark the dimensions of the tasks which could be solved by human beings with relatively small number of errors. Even for such tasks they need help to find and eliminate the errors. For the problems of bigger size the necessity of the help is evident. That why it is reasonable to solve the multicriteria classification problems with the help of the method ORCLASS.

Interactive system ORCLASS and its implementation

On the basis of the described approach a decision support system ORCLASS for ordinal classification tasks was developed.

ORCLASS is a system which helps a decision maker (user) to construct and use classification rules for identification of appropriate classes for objects estimated upon a set of criteria. The system supports three main tasks throughout the process of task solution:

- entering/editing initial data (alternatives, attributes and their scales, estimation of alternatives against attributes, decision classes);

- construction of the complete and contradiction free classification of all possible multicrtieria objects on the basis of the decision-maker's (user's) preferences (through a flexible dialogue with the user);

- elaboration of classification rules on the basis of the constructed classification, their analysis and modification (if needed), and implementation of classification rules for identification of appropriate classes for multicriteria objects.

The decision support system ORCLASS is a user-friendly tool. The system allows to use natural-like language for evaluation of multicriteria alternatives (through implementation of verbal estimations of criterion' values and verbal formulation of decision classes). The system gives the opportunity to detect and eliminate possible inconsistencies in the user's responses, provide explanations of the result. The dialogue is organized on the basis of psychologically valid procedures.

The system can be used in many real situations, where it is necessary to identify appropriate decision class for the object under classification: make a decision about an article, submitted to a scientific journal; to diagnose a patient, to find the fault in a technical system and so on.

First, the system of criteria is elaborated, and appropriate decision classes are formulated. This process is not formalized and is carried out prior to the computer use.

Second, the construction of the complete classification for all possible multicriteria objects is carried out according to the method ORCLASS described earlier.

The initial information necessary to start the work with the system consists of criteria with their scales, and a list with classes of decision (see criteria and decision classes for evaluation of articles submitted to the scientific journal). As it has been stated above, all criteria have ordinal scales and verbal descriptions of quality grades. Decision classes are also rank ordered from the best to the worst and also have verbal descriptions. Number of criteria (Q) is limited to 10, and number of criterion values for one scale is limited to 9.

As the number of criteria and number of criterion values are finite and discrete, all hypothetical combinations of criterion values are formed by a Cartesian product of scales. The system is restricted to number of elements in Cartesian product (capacity of the set Y) equal to 5000 (it means the system is able to work with not more than five thousand possible vectors in the set Y).

The system allows to construct the complete and contradiction free classification of all elements from the set Y by presenting part of these vectors to the decision maker for classification.

Construction of the complete and contradiction free classification of multicriteria objects

The system ORCLASS forms the set of all possible vectors of the set Y and for each of them it defines the set of permissible classes G_i (i=1,2,...,N). At the first step these sets contain all decision classes (from 1 to L) for all elements from Y, but two: element y_1 from Y with all the most preferable criterion values is assigned class I ($G_1=\{1\}$), and the vector y_N from Y with all the least preferable criterion values is assigned class L ($G_N=\{L\}$). Thus, $G_i = \{1,2,...,L\}$, i=2,3,...,N-1.

In accordance with this information the system ORCLASS calculates the "informative" index for each vector from Y (according to the expression (5.4)).

Then the vector y_i with the maximum index value is selected and displayed on the screen for classification by the decision maker. An example of such presentation is given in figure 5.6.

If the response of the decision maker (class for the presented alternative) is within the limits of the corresponding set G_i, the response is considered to be consistent with the previous information and is stored in G_i. In accordance with the received information the sets G_j are modified for all y_j, which dominate or are dominated by the vector y_i.

Figure 5.6. Visualization of the hypothetical alternative for classification and a menu of possible responses

---

THE FOLLOWING SITUATION IS UNDER CONSIDERATION:

1. The article has correspondence to the journal's outline.
2. The results have some theoretical value in the field.
3. The results have practical value.
4. There are essential errors and inconsistencies in the article.
5. The article is not nicely written.

POSSIBLE ANSWERS:

Class I.   Article is to be published.
Class II.   Article has to be revised and then published.
Class III.   Article has to be revised and then reviewed again.
Class IV.   Article is to be rejected.

YOUR ANSWER:

After that the system checks if there exist unclassified elements. If all elements of the set Y are classified, it is concluded that the complete and contradiction free classification of all elements from Y is constructed.

If there are unclassified elements in the set Y, the informative indices for them are recalculated, and an element with the maximum value of this index is presented to the decision maker for classification.

If the decision maker's response is not within the limit of permissible classes, it is concluded that this response contradicts to the previous ones, and the system ORCLASS informs the decision maker about this fact and suggests to alter his (or her) last response, or to analyze the situation. If the user prefers not to change the last response, but to analyze the contradiction, the system displays relevant information as shown in figure 5.7.

It is assumed that the decision maker will analyze the situation and resolve the conflict by modifying the class for one (or both) alternatives. If the last response of the decision maker was inconsistent not only to one but to larger number of other responses, and the decision maker did not change it, the system will present in pairs all elements of the set Y_{err} (this set has all vectors which previous classifications contradict to the last decision maker's response), to guarantee that the decision maker wants to change all previous responses, inconsistent with the last one.

Figure 5.7. Visualization of inconsistent responses with explanations

---

ANALYSIS OF INCONSISTENT RESPONSES

1. The article directly corresponds to the journal's outline.
2. The results have theoretical value in the field.
3. The results are of high practical value.
4. There are essential errors and inconsistencies in the article.
5. The article is written awfully.

THE SITUAION IS CLASSIFIED AS:
Class II: The article is to be revised and then published.

1. The article directly corresponds to the journal's outline.
2. The results have theoretical value in the field.
3. The results are of high practical value.
4. There are essential errors and inconsistencies in the article.
5. The article is not nicely written.

THE SITUATION IS CLASSIFIED AS:
Class III: The article has to be revised and then reviewed again.

> The second situation is more preferable than the first one according to the criterion values. It must be assigned to a not less preferable class than the first situation. Analyze the inconsistency and classify both situations again.
> PRESS ANY KEY TO CONTINUE

After that the system will check if introduced modifications in classification of vectors from Y caused the appearance of new contradictions (see procedures and algorithms above), and will work on them with the decision maker to construct the consistent classification. After that the system will propagate these new classifications on the basis of dominance relation to modify correspondingly sets G_j.

The interview is continued up to the moment when the classification is built. After that, the system provides the user with the possibilities to analyze the classification rules, used by him (or her) in this task.

Introduction and modification of classification rules in the system ORCLASS

The construction of the complete and contradiction free classification of the set Y provides all the necessary information for identification of appropriate class for any possible object, described using the formed set of criteria. At the same time to give explanations for the made decisions and to be able to verify the introduced system of rules for classification, the classification is presented as a number of boundary elements for each class (see figure 5.4). This presentation gives the decision maker the possibility to analyze the constructed boundaries. If the decision maker is not satisfied with any of them, it is possible to modify them in accordance with the decision maker's preferences.

Let assume the decision maker used the constructed classification for identification of appropriate classes for a number of elements (e.g., articles submitted to the journal). The result was that all the articles were assigned the last (least preferable) class. This result may lead the decision maker to the desire to decrease the level of requirements for the submitted manuscripts (against some of the criteria).

The system allows to carry out the modification of the obtained boundaries. To do this, the mode of introducing the specific vector from Y for classification is used. In this mode, the decision maker is able to enter a vector from Y, which he (or she) wants to classify (or reclassify). In this mode it is possible to enter even classified vector (for reclassification). After entering this vector, the system presents it for classification in the common way (see figure 5.6).

The system will input the new decision maker's response and will work in a standard regime. If the response contradicts the previous one(s) (as will be the case if we want to change the class for the boundary element), the system will help to eliminate contradictions and to form modified classification.

Thus, introducing one (or several) element(s) of upper or/and lower boundaries, the decision maker initiates the work of reclassification of all vectors, which classes contradict to the newly introduced ones. As a result, boundaries are reanalyzed and modified in accordance with the decision maker's preferences.

This system allows to modify classification rules in a quick and accurate way. This mode also allows to speed the process of the construction of the complete classification by introduction of rules, which the decision maker is able easily formulate (if any). In any moment of working with the system ORCLASS, it gives the user the possibility to enter vector for classification, providing more steady and effective way of selecting the vector with the maximum informative index.

Final boundaries obtained through this interactive process (which presents classification rules as was stated above), are used for classification of real objects. In this case, the user is to enter into the system the appropriate alternatives with corresponding values upon the formed set of criteria. The alternatives are presented in the system ORCLASS as corresponding vector from Y. The system ORCLASS finds the assigned class for the vector from Y, which describes the considered alternative and presents it to the user with the generated explanation (see figure 5.8).

Figure 6.8. An example of an explanation in the system ORCLASS

Alternative A1 (vector 22211), which is described as:

1. The article has correspondence to the journal's outline.
2. The results have some theoretical value in the field.
3. The results have practical value.
4. There are no errors and inconsistencies in the article.
5. The article is nicely written.

is assigned : CLASS I - THE ARTICLE IS TO BE PUBLISHED,

as in the lower boundary of this class there is vector 23311, which is less preferable than alternative A1 against criteria 2 and 3.

Criterion 2. Theoretical value of the results
Criterion 3. Practical value of the results

PRESS ANY KEY TO CONTINUE

When working with the system the user is able to get all the information about the data presented in the system, and the level of task solution. The data in the system include: criteria and their scales, decision classes, real alternatives, alternatives with the appropriate decision classes (it this stage of work has been carried out), class boundaries (if the complete classification has been built).

The level of the task solution informs the decision maker about the current step in the decision process. If the complete classification is built the system informs the decision maker about that and supplies information if real alternatives are classified.

If the complete classification is not constructed, the system informs the user about the number of elements in the Y classified and left unclassified up to the moment. Upon the request of the user, the system is able to display all vectors classified by the user. This mode may be used to check, or change the previous classification of vectors (if the user knows what he, or she wants to modify).

The system requires about 200 Kbytes and can be run on IBM PC compatibles with MS DOS 3.3 or higher.

Implementation of the system ORCLASS for evaluation of R&D projects in a medical research institution

Traditionally in Russia medical research is largely concentrated in special research institutions fully supported by the state financing. A lot of money during last decades were spent on research in the area of cancer diseases. One of the main centers for the research is Moscow Oncological Research Institute, which has hundreds of researchers and obtain advanced facilities for the research.

During the period of perestroika and glasnost, and economical difficulties of the time, the level of the state support for majority of research activities were reduced, at the same time the more comprehensive rights of the organizations for selection of R&D projects to be supported by the resources at hand were granted. These condition led to the decision of the Institute administration to elaborate an effective on-line system of R&D proposals' evaluation.

Traditionally used system of multicriteria evaluation of R&D projects, with quantitative measurements and importance weights to achieve the overall score for each proposals was considered to be inappropriate in new conditions, when the decisions were to be made on a regular basis, within the limited resources, and had to be explained to the applicants as well as to the higher authorities in conflict situations.

First, it was decided that it is necessary for each submitted proposal to be assigned an overall assessment group, considering the overall level of the project. Projects of the first priority group were to be fully supplied with the necessary resources. Other groups were to be supplied at a lesser level (considering the resources at hand).

Next, it was realized that characteristics (criteria) which are to be used for evaluating of R&D projects were mostly of qualitative nature. It would be difficult to use numerical form of their assessment by different experts (as R&D projects were to be evaluated by experts against criteria). There would be the possibility of biases in their estimates (due to individual perception of adequacy between verbal and numerical values for them).

That's why it was decided to use method (and system) ORCLASS to solve the task, which works with verbal criteria, and helps to divide all possible alternatives among small number of decision classes. Consultants worked together with specialists from the Institute to construct the required system for R&D projects' evaluation.

To provide the feedback between the proposed research and the obtained results (as planning was carried out on the annual basis), it was also decided, that there would be too parallel systems: the aim of the first system was to evaluate proposals, and the aim of the second system was to evaluate the results of the supported proposals (Larichev et al., 1990).

The first step recommended by consultants was to elaborate the system of criteria for evaluation of R&D projects and results. That set of criteria was to be used by experts while evaluating R&D projects, and the same set was to be used by the decision maker (Coordinating committee of the research center in this case) while elaborating the classification rules for overall assessment of alternatives. Along with the set of criteria, the set of decision classes was elaborated.

CRITERIA FOR EVALUATION OF R&D PROPOSALS

Criterion A. Urgency of the research
A1. The problem is of high priority today
A2. The problem may be important
A3. There is such a problem
A4. The problem is not important today

Criterion B. Level of the proposed research
B1. The research is of high level inside the country and abroad
B2. The research is of high level inside the country
B3. The research if not of high level inside the country

Criterion C. Prospects for the research
C1. The results can introduce principal changes into the activity of the cancer service and/or the medicine development
C2. The results can essentially influence the activity of the cancer service and/or the medicine development
C3. The results can be of small influence on the activity of the cancer service and/or the medicine development

Criterion D. Type of the result

   (most common types of implementation of the results)

D1. Normative documents with detailed description of methodology, or medical supplies, or medical devices

D2. Instruction materials, technical innovations, cooperative research programs

D3. Scientific and information materials, publications

D4. Materials for educational programs

Criterion E. Level of implementation

E1. The state level

E2. The republican level

E3. Regional level

E4. Local level

Criterion F. Level of resources' supply

F1. The research is supplied with the resources at a high level

F2. The research is supplied with the majority of necessary resources

F3. The research is supplied with some resources

F4. The research has almost no resources.

DECISION CLASSES

Class I:	The project is of high priority
Class II:	The project is of middle priority
Class III:	The project is of low priority
Class IV:	The project is unsatisfactory.

To evaluate the results of the projects, the same criteria were used except for the last one (criterion F), which was changed for "The level of protection, provided by the results"), and the criterion E was reformulated (for all projects except experimental ones) as "Effectiveness of the results". The decision classes were the same.

The dimensions of the task under consideration are: number of criteria (Q) is equal to 6; number of criterion values (n_q) is equal to 4 for criteria A, D, E, and F, and is equal to 3 for criteria B and C; the number of all possible combinations of criterion values (the capacity of the set Y) is equal to 2304. The dimension is rather large, so the effective procedure of the dialogue with the decision maker, while constructing the classification rules, based on the presentation of informative vectors from Y, was a necessity.

An additional problem arised concerning the "impossible" combinations of criterion values. While selecting the most informative vectors for classification by the decision maker, the system ORCLASS in a formal way selects one of the vectors from the set Y. At the same time, when some of these vectors were presented to the decision maker, the response was that it was not reasonable to classify such a vector, as such combination of criterion values was impossible in real practice. For example, if the R&D project was considered to be "not important at the time" (criterion value A4), then the results of the project could not lead to "principle changes in the activity of the cancer service, and/or medicine development" (criterion value C1).

To overcome this problem, the system was modified to be able to state special class "impossible situation" among decision classes. In this case, the combination of criterion values, which were considered to be impossible in real practice by the decision maker, were entered into the system, and all vectors in the set Y with such combinations were marked as the "impossible" and were not used further in the dialogue with the decision maker.

The decision maker in this task (while working with the system ORCLASS) was presented by an expert group of the Coordinating Committee (three members). That's why the results of working with the system (impossible combinations, class boundaries, and corresponding production rules) were discussed several times at the Coordinating committee meetings to work out the necessary modifications and to adopt the final version of classification rules.

The results of the first session working with the system were used mainly to work out the "impossible combinations of criterion values". These combinations were presented to the Coordinating committee and were discussed. Some changes were made. As a result, about 16 impossible combinations were accepted, and were used while constructing the classification rules. They covered almost half of all possible vectors from the set Y.

The process of elaboration of the final classification rules was carried out through several stages. The expert group worked out the variant of classification rules, they were analyzed and presented to the Coordinating committee. The committee analyzed them and decided on the necessary changes in the boundary elements (or just reformulating some of the rules, or introducing new ones). After that the expert group returned to the system and introduced all these modifications, using the mode of the modification of the classification rules (described in previous section of this chapter). Their task was also to work out all the appeared inconsistencies with the previous responses, and to classify some of vectors which turned out to be unclassified due to introduced modifications. The results were analyzed and presented to the Coordinating committee once again, and so on. It was necessary to carry out several such steps for the committee to accept the final set of classification rules (see figure 5.9).

These classification rules divide all vectors from the set Y among decision classes in approximately following proportions: 20% of all vectors belong to class I; 32% of all vectors belong to class II; 20% of all vectors belong to class III, and 28% of all vectors belong to class IV.

Figure 5.9. Class boundaries for the final variant of classification rules for evaluation of R&D projects

CLASS I
Elements of the upper boundary of the class: 111111
Elements in the lower boundary of the class:
322322 222232 232222 111432
CLASS II
Elements of the upper boundary of the class: 111113
131131 131311 213211 311131 112331 121331 211331 212141
Elements of the lower boundary of the class:
322443 332323 232233 323322
CLASS III
Elements of the upper boundary of the class:
111114 131331 213231 213213 233211 331131 232141
Elements of the lower boundary of the class: 332443 223441
232324 322324 332224   223233 223323 22332 233322 323223
323232
CLASS IV
Elements of the upper boundary of the class: 111134 313313 213342 213214
313331213432 333211 413321 233313 233231 213333 313323 331314
Elements of the upper boundary of the class: 433444

The presented class boundaries demonstrate rather complicated set of classification rules for evaluation of R&D projects, when some increase (decrease) in the quality against one criterion may change the appropriate class for the project under consideration. For example, the R&D project may be considered of high priority (class I) if it has at least main resources for the project (criterion value F2) and has "essential influence on the activity of cancer service and/or medicine development" (criterion value C2). At the same time the results are to have not less than the second values against all other criteria but B or E, or A and D together (it can have the third values against these criteria). The project can be considered of high priority even if it has not good innovation level (criterion value D3), but in this case it has to have all the best criterion values against criteria A, B , and C.

Different class boundaries were obtained for the case of evaluating <u>results for supported projects</u> (see figure 5.10).

Figure 5.10. Class boundaries for the final variant of classification rules for evaluation of the results for supported R&D projects

---

CLASS I

Elements of the upper boundary of the class: 111111

Elements in the lower boundary of the class: 323421 323323 223422
233223 233322 322233 322323 322332 323223 323322

CLASS II

Elements of the upper boundary of the class: 213131
221333 333122 313422 223423 233323 233422 323232

Elements of the lower boundary of the class: 323423
333422 233323 322333 333223 223231 313231

CLASS III

Elements of the upper boundary of the class:
213132213331 323131 233423 333323

Elements of the lower boundary of the class:
333423323431 223332 223233 323232

CLASS IV

Elements of the upper boundary of the class: 233132
213432 313332 223333 323133

Elements of the upper boundary of the class: 433444

In this case the distribution of vectors from the set Y is as follows: 86% of all vectors belong to class I, 7% of all vectors belong to class II, 4% of all vectors belong to class III, and 3% of all vectors belong to class IV. The observed uneven distribution was analyzed, and it was concluded that the more positive evaluation of different alternatives in this case is connected with the fact that obtained (known) results are evaluated. In case of proposed R&D projects it is known that expected results are usually of much higher level and significance than those, that are really obtained in many cases. Positive results of the research are enough to consider that the resources are not spent in vain.

The carried out implementation of the method and system ORCLASS showed several positive characteristics connected with this approach. First, all the members of the Coordinating Committee agreed that the set of so detailed and complicated classification rules could not be elaborated without the help of the system. At first steps of the work, there were several attempts of the Coordinating Committee to form a set of rules for classification. When entered into the system ORCLASS (using mode of the introduction of the vector for classification), they were shown to cover only part of all possible vectors from the set Y (from 40 to 70 percent), and in some cases were inconsistent.

Second, the Coordinating Committee considered this approach to be helpful while dealing with the elaboration of the decision rules by a group of decision makers (committee itself). Easy and complete presentation of all rules at each iteration in the process of rules' development, allowed the committee to deal effectively with diversity in opinions and to come to many agreements more quickly and in a more finite (and understandable) way. It was much easier for the committee members to agree on classification of a specific combination of criterion values, than on more general rule formulation.

All these show that the approach (as well as the system) can be used effectively in many real decision situations.

Conclusion

The described method and system are meant for the solution of ordinal classification tasks, based on the possibility of a decision maker to classify separate alternatives. To our mind these tasks are wide spread in practice but not very popular with specialists in decision-making. System ORCLASS allows one to elicit information (or knowledge) in a traditional form (through qualitative criteria scales and verbal descriptions of decision classes). It provides the possibility to receive reliable information as it tests this information for consistency. ORCLASS generates a complete classification rule, which makes it possible to find the decision class for any combination of criterion values, reducing the number of questions by choosing the most informative ones.

The system ORCLASS allows a user friendly interface and does not require special computer skills. At the same time, the method itself requires adequate knowledge in the area of multicriteria decision making. That's why the system is meant mainly for the consultants working for the decision makers..

This system has been used in a wide range of practical tasks, from R&D planning context (Larichev and Moshkovich 1986, 1987, 1990; 1992; Larichev et al., 1990) to medical diagnostics (Larichev et al., 1986, 1989, 1990, 1991).

CONCLUSION

Scientific validity of verbal decision analysis

Scientific value of theories is known to be defined by their ability to forecast correctly one or another phenomenon. Yet, if we apply this criterion to the decision theory, then we will face intricate procedural problems. In contrast to economics and operations research, the models of decision making are subjective and reflect the DM's vision of the problem under study. This subjective concept is represented in form of a set of criteria, formulations of estimates, and composition of alternatives.

It is apparent that a correctly made decision can turn out unsuccessful in the course of subsequent events that are out of DM's control. On the contrary, a decision, which is unsuccessful from the current point of view, can quite unexpectedly prove to be a success as a result of a development of events.

It is, therefore, obvious that in order to validate a decision method the DM must make a fairly great number of decisions. A sufficiently long time is required for their consequences to manifest themselves and results to be assessed objectively.

This approach was used by the present authors to validate the ZAPROS method (Ch. 3) which was used by the centralized system for evaluation of applied R and D projects over a five-year period (Zuev et al., 1979). Each project involved some indefiniteness. Independent experts evaluated each of them at the stage of planning by six criteria on verbal estimate scales shown in Ch. 3. The DM's preferences underlie the construction of a quasiorder on the set of projects (see Ch. 3). The unsatisfactory projects were rejected.

It is only natural that among the accepted projects there were better, fair, and worse ones. After the five years of work, each project was estimated as successful or not by *ad hoc* committees using a multicriteria scheme with other verbal criteria. The ZAPROS method again was used to construct the decision rule from the preferences of the same DM.

Together, 750 applied R and D projects were estimated. The prognostic power of ZAPROS could be validated from its success in predicting success or failure of one or another project at the stage of planning. To do so, the accepted projects in quasiranking were divided with the help of DM into three groups of best, fair, and satisfactory projects. The same was done for the quasiranking of the resulting actual estimates of projects. Analysis demonstrated that on the set of 750 projects the correlation was 82% (Chuev et al., 1983). In our view, this is a good result for a large sample of typical nonstructured problems.

DM's personality and decision rule influenced, of course, the result of comparison, as well as the accuracy of expert estimates influenced the prediction of project results. At the same time, the results of each project were affected by many unpredictable factors such as competitive results obtained in other countries, unexpected difficulties, etc.

A great number of projects and sufficiently long time for estimating the consequences of decisions provide the ground for `external' estimation of one or another decision method.

Yet, there are no less important `internal' criteria. It is possible and necessary to estimate the reliability of procedures for preference elicitation. For example, when comparing objects differing in estimates by two criteria, near the so-called reference situation (see Ch. 3) with 6 criteria having 3 verbal estimates on scales (25 through 45 comparisons) people have on the average 2 or 3 contradictions. Chapter 4 presents the results of classification of multicriteria alternatives by the ORCLASS method. For problem size as defined in domain of human possibilities (Ch. 4), there are from 3 to 6 contradictions. In both methods, means are provided for detecting contradictions and presenting them to the DM for analysis.

Thus, we are able to predict the degree of DM's infallibility in man-computer interaction. This prediction demonstrates the practical value of one or another method of elicitation of preferences.

Another internal criterion is low sensitivity to human measurement errors. The validity of a decision method is also ensured to some extent by such its features as checks of criteria for consistency and preference-independence and possibility for DMs to learn from their errors and offer explanations.

We believe that the scientific value of the prescriptive decision methods must be determined using both internal and external criteria.

Role of theory and practice

A consultant comes to the DM's study, and they begin their first conference. What are they discussing? It is only natural that they touch upon matters important for the DM. The range of nonstructured problems can be extremely wide --- growth of social discontent, intention of provinces to separate from the central authorities, behavior of market competitors, diversification of products, etc. But it is important that in all cases they speak a language which is customary and understandable to the DM.

It is difficult to imagine what 'wonders' must be promised to DMs to make them master the fundamentals of the subjective utility theory or compare lotteries. We can assert from our practice that such proposal of the consultant would make their fist meeting the last, because, if analysis of decisions is close to clinical analysis as was first indicated by B. Fischhoff (Fischhoff, 1983), one cannot imagine a patient studying medicine and Freudian psychoanalysis before the physician starts to treat him/her.

How should the consultants act to be successful and interesting to the collocutor, get a good order, and render real help to DMs? They must be experienced, efficient, and clever. We cannot cite a textbook of this art --- the art of primary structuring of problems, the art of finding their 'roots' (Checkland, 1981) --- and this is not by accident because such art cannot be expounded in writing, but is taught in real life through concrete examples, like diagnosis of diseases is taught in medicine.

We note that there are very good books on operations research (Rivett, Ackoff, 1963) which provide an insight into the intricacy of the initial analysis of problems. Like the beginning physicians, the consultants from year to year master their art by imitating their more experienced colleagues, by succeeding and erring. But when the first stage is passed, the consultants have to demonstrate to the DMs that, apart from rapid understanding of the difficulties of other persons, they have certain professional background. They must render real help to the DMs by asking them unexpected questions in their language and helping them to reach a difficult compromise between contradictory estimates and conflicting active groups.

The verbal analysis of decisions, whose fundamentals are presented in this book, is in our view one of the instruments in the consultant's toolkit. But we again stress that this instrument is neither unique, nor or universal. It might be well to note that the peculiarities of the methods of verbal analysis of decisions make them 'transparent' to the DMs. This fact sometimes proves to be of advantage to the consultant.

Therefore, if primary analysis, search of approach, and structuring of problem are an art, then this art is subconscious like that of physician or chessplayer (Kihlstrom, 1987 , Lewicki et al., 1992). Yet, we are well aware of the fact that there exist numerous books about openings and ends in chess or medical textbooks and monographs. Stated differently, there are means of transmitting the declarative knowledge. And books on decision making, including also this book, serve the same purpose by offering practical methods to the consultants.

Is it possible to separate in practice the decision methods and the art of problem analysis for unstructured problems ? Is it desirable, being a good professional in analyzing nonstructured problems, to make use of quantitative methods of measurement and comparison accompanied by the so-called sensitivity analysis? We do not think that this is the best way of action, which is corroborated by the case-study of systems analysis (see Ch. 1). The cost-effectiveness method has amalgamated with structuring (system approach) and made up an organic whole. The famous PPBS system (Novick,1965), requires that a program structure group departmental activity so as to enable comparison of cost and effectiveness of alternative approaches to reaching the department's goal and that program memoranda show why certain courses of action were chosen by formulating the department objectives in a measurable form.

The criticism that followed the first advances of the systems analysis (Hoos, 1972) was directed to the quantitative indicators and requirements that social goals be represented in a purely quantitative form. Failures in using the systems analysis were mostly due precisely to the methods of comparison of alternatives accepted within its framework.

What is in the DMs' minds?

The researchers of decision making were actively discussing over the recent years the question of information stored in the DMs' minds. What can we anticipate by posing to the DMs questions following from the logic of one or another decision method? As was noted in the well-known book of (Winterfeldt, Edwards, 1986), it can be hardly expected that the utilities and numbers expressing the subjective estimates of objects and situations are stored there. We can only agree with the authors that it is more likely to assume that the DMs have ordinal relationships between parameters (ordering of criteria, estimates on scales, etc.).

The experience that we gained in working with DMs on nonstructured problems enables us to formulate the following remarks:

1. Only a part of critical criteria are used by the DMs in making their decisions. Usually, they are aware of all substantial criteria for alternatives' evaluation, but make use of one or another group depending on circumstances. The first aid which the consultant can render to DMs is to systematize the composition of criteria and develop scales of estimates.

2. Only the ordinal relationships of main elements of DMs' policy are understood by them sufficiently clearly, coherently, and consistently. Stated differently, there are criteria and also some estimates on their scales where the DMs are always consistent. Yet, for less important criteria they can make inconsistent comparisons of importance or inconsistent comparisons of

estimates. We can confidently claim that in nonstructured problems the main elements of policy are well understood by the DMs, but its details are not elaborated. Consultants and computer systems must help them to rework and specify their policy .

3. The DMs are well aware of their own problems which can be called the main goals of their activity, but they are often limited by the existing initial set of alternatives and see no way to extend it. Conditions are required to foster new ideas and new views of the problem. It is unlikely that such ideas can be obtained from the consultants or computer systems. They can come only from experts representing a wider circle of professionals; but they need to be asked correctly and their quest must be vectored in the necessary direction. The aid which the consultants and computer systems can render to DM consists in formulating the requirements on new desirable alternatives.

Directions of future research

We would like to conclude the book by emphasizing once again how amazingly diverse is the world of human decisions. It can accommodate all existing approaches. For many problems, methods of their analysis have not yet been created.

In our judgment, the decision making as a field of research is in the making --- transition from universal approaches and methods of decision making to methods oriented to certain classes of problems is looming ahead. Stated differently, transition from seeking panaceas for all problems to careful analysis of problem features and determination of adequate decision methods is under way. More and more works are devoted to comparing decision methods. We hope that their number will be increased in future.

Decision making is often regarded as one of the areas of economics. Indeed, problems of decision making occur both in microeconomics and in macroeconomics. Although decision making does not address exclusively the economic problems, the latter are met very often. From the point of view of economics as science, the approach of decision making has basic distinctions. From elaborating the models of economical systems ,describing the behavior of particular consumers and producers in a particular environment we pass to particular means enabling people to make sensible decisions.

We believe that science must not only explain but help to act. In this book we tried to show how to analyze variants of decisions for problems described in qualitative terms.

REFERENCES

Aho A.V., Hopcroft J. E., Ullman J. D., (1974). *The Design and Analysis Of Computer Algorithms*, Addison-Wesley, Reading, MA.

Aizerman M. A., Aleskerov F., (1990). *Choice of variants (foundations of the theory)*, Moscow, Fizmatlit Publ.. (in Russian)

Andreeva Ye., Larichev O. I., Flanders N. E., Brown R. V., (1995). *Polar Geography and Geology*, 19, 1, 22-35.

Atkinson R. C., Herrmann D. J., Wescourt K. T., (1974). Search processes in recognition memory, In: R. L. Solso (*Ed.*), *Theories in Cognitive Psychology*: The Loyola Symposium, Hillsdale, New Jersey, Erlbaum Ass., pp. 101-146.

Atkinson R. L., Atkinson R. C., Smith E. E., Bem D. J, (1993). *Introduction to Psychology*, 12-th edition, Harcourt Brace Jovanovich Publ.

Baddeley A., (1994). The Magical Number Seven: Still Magic After All These Years?, *Psychological Review*, V 101, N 2, pp. 353-356.

Bana e Costa C. A., Vansnick J. C., (1994). The MACBETH approach : general overview and application, *Centre of urban and regional systems of the Technical Unversity of Lisbon*.

Bana e Costa C. A.., (1992). Absolute aand relative evaluation problematics. The concept of neutral level and the MCDA Robot technique. In: Cerny, D.Gluckaufova, D.Loula

(*Eds.*) *Proceedings of the International Workshop on Multicriteria Decision Making. Methods-Algorithms-Applications. Liblice, 18-22 March, Prague.*

Berkeley D., Humphreys P., Larichev O.I., and Moshkovich H.M., (1990). Modelling and supporting the process of choice between alternatives: the focus of ASTRIDA. In: H.G. Sol and J. Vecsenyi (*Eds.*), *Environments for supporting decision processes.* Amsterdam: North Holland, 59-62.

Bertier P., Bouroche J., (1975). *Analyse des donnees multidemensionnelles.*, Paris, Presses Universitaires de France. (in French)

Bonczek R. H., Hoffsapple C. W., Whinston F. B., (1981). *The evolution from MIS to DSS.*, New York, Academic Press.

Borcherding K., Schmeer S., Weber M., (1993). Biases in multiattribute weight elicitation, In: J.-P. Caverni, M. Bar-Hillel, F. N. Barron, H. Jungermann (Eds.), *Contributions to Decision Research.*, North-Holland.

Borcherding K., von Winterfeldt D., (1988). The effect of varying value trees on multiattribute evaluations, *Acta Psychologica*, 68, pp. 153-170.

Brown R. V., Kahr A. S., Peterson E., (1974). *Decision analysis for the manager*, New York, Holt, Rinehart and Winston.

Budescu D., Wallsten T., (1995). Processing linguistic probabilities :General Principles and empirical evidence, *The Psychology of Learning and Motivation*, v. 32, Academic Press.

Carnap R.,(1969) *Philisophical foundation of physics.*, London, Basic Books Inc.Publishers.

Chance N. A., Andreeva Ye. N., (1995). Sustainability, equity, and natural resource development in Northwest Siberia and Arctic Alaska, *Human Ecology*, 23, 2, 217-240.

Checkland P. B., (1981). *Systems Thinking, Systems Practice.* New York, Wiley.

Chuev Ju. V., Larichev O. I., Zuev Ju. A., Gnedenko L. S., Tichonov I. P., (1983). Interactive procedure for planning of R and D projects, In: D. Chereshkin (*Eds.*) *Problems of Information Technology*, Proceeding of VNIISI, N 6, pp. 86-95. (in Russian).

Clarckson G. P., (1962). *Portfolio selection: A Simulation of Trust Investment.* N.Y.: Prentice Hall, Englewood Cliffs.

Cook W. D., Kress M. A., (1991). Multiple Criteria Decision Model with Ordinal Preference Data, *European Journal of Operational Research*, 54, 191-198.

Cordier M., (1984). Les systems experts, *La Recherche*, v. 15, (in French)

Currim I. S., Sarin R. K. (1989). Prospect versus Utility, *Management Science*, V. 35, N 1, Jan., pp. 22-41.

David H.A., (1988). *The method of paired comparisons*, Second Edition. Oxford Univ, Press: USA.

Dawes R. M., (1982). The robust beauty of improper linear models in decision making, In : D. Kaheman, P. Slovic, A. Tversky (*Eds.*), *Judgement under uncertainty: Heuristics and Biases*, Cambridge, Cambridge Univ. Press.

Dawes R. M., (1988). *Rational choice in an uncertain world*, Harcourt Brace Jovanovich Publishers, N. Y.

Dolbear F. I., Lave L. B., (1967). Inconsistent behavior in lottery choice experiments, *Behaviour Science*, v. 12, N 1.

Dror Y., (1989). *Public Policy Making Reexamined*, Transaction Publishers, Oxford.

Enthoven A., (1969). The systems analysis approach. In: H. Hinrichs, G. Taylor *(Eds.)* *Programm Budgeting and Benefit-cost Analysis: cases, text and readings*, Goodyear Publ. Comp. Inc., California, Pacific Palisades.

Erev I., Bornstein G., Wallsten T., (1993). The Negative Effect of Probability Assessments on Decision Quality, *Organizational Behavior and Human Decision Processes*, 55, 87-94.

Erev I., Cohen B., (1990). Verbal versus Numerical Probabilities: Efficiency, Biases, and the Preference Paradox, *Organizational Behavior and Human Decision Processes*, 45, pp. 1-18.

Erev I., Wallsten T., Neal M., (1991). Vagueness, Ambiguity, and the cost of mutual understanding, *Psychological Science*, v 2, N 5.

Farquhar P., Pratkanis A., (1991). Decision Structuring with Phantom Alternatives, *Carnegie Mellon University Report*.

Fischhoff B., (1996). The Real World: What Good is it?, *Organizational Behavior and Human Decision Processes*, v 65, N 3, pp. 232-248.

Fischhoff B.,(1983) Decision Analysis: Clinical art or clinical science?,In: L.Sjoberg, T.Tyszka, J.Wise (Eds.) *Human Decision Making* ,Doxa,Bodarfors,Sweden.

Fischhoff B., Goltein B., Shapira Z., (1980). The experienced utility of expected utility approaches, *Tech. Rep. N PTR -1091-80-4*, Eugene, OR, Decision Research.

Fischer G.(1991) *Range Sensitivity of Attribute Weights in Multiattribute Utility Assesement.*,Duke University.

Fishburn P. C., (1970). *Utility theory for Decision Making*. New York: Wiley.

Furems E. M., Moshkovich H. M, (1984). Ordering of vectors for task of portfolio selection, In: O. I. Larichev *(Ed.)*, *Procedures for estimating multiattribute alternatives*, VNIISI press, Moscow. (in Russian).

Furems E. M., Vasunin G. N., Larichev O. I., Chernov Yu. Ja.., (1982). Problems of bin packing with multiple criteria. In: S. V. Emeljanov, O. I. Larichev *(Eds.)*, *Problems and methods of decision making in organozations*. Moscow: VNIISI press, 84-91. (in Russian).

Gallhofer I.N.,SarisW.E. (1989) Decision trees and decision rules in politics:The empirical decision analysis procedure.,In: O.Svenson,H.Montgomery (Eds.) *Processes and structure in human decision making.*,New York,Wiley.

Garey M. R., Johnson D. S.,(1979). *Computers and Intractability*, W. Freeman and Co., San Francisco.

Gnedenko L. S., Larichev O. I., Moshkovich H. M., Furems E. M., (1986). Procedure for Construction of a Quasi-Order on the Set of Multiattribute Alternatives on the Basis of Reliable Information About the Decision-Maker's Preferences. *Automatica i Telemechanica*, 9, 104-113. (in Russian)

Gonzalez-Vallejo C., Wallsten T., (1992). The effects of communication mode on preference reversal and decision quality, *Journal of Experimental Psychology: Learning, Memory and Cognition,* 18, pp. 855-864.

Goodwin P., Wright G., (1991). *Decision analysis for management judgments.* N.Y.: Wiley.

Hamm R.., (1991). Selection of Verbal Probabilities Solution for Some Problems of Verbal Probability Expression, *Organizational Behavior and Human Decision Processes,* 48, pp 193-223,.

Harte J. M., Westenberg M. R., van Someren M., (1994). Process models of decision making, *Acta Psychologica,* v. 87, , pp. 95-120.

Hayes -Roth F., Waterman D., Lenat D., (*Eds.*) (1983). *Building Expert Systems,* London, Addison-Wesley Publishing Co.

Hinrichs H., Taylor G. (*Eds.*), (1969). *Program Budgetingh and benefit-cost analysis : cases, text and readings,* Goodyear Publ. Company, Inc., Pacific Palisades, California.

Hoos I. R., (1972). *System Analysis in Public Policy,* University California Press, Los Angeles.

Houston M. C., Ogawa G., (1966). Observations on the theoretic bases of cost-effectiveness, *Operations Research,* v. 14, N 2.

Huber B., Huber O., (1987). Development of the Concept of Comparative Subjective Probability, *Journal of Experimental Child Psychology,* 44, 304-316.

Humphreys P. C., (1977). Application of multiattribute utility theory, In: H. Jungermann, G. De Zeeuw (*Eds.*), *Decision making and change in human affairs,* Dordrech, Reidel.

Humphreys P. C., Wishudha A., (1979). Multi Attribute Utility Decomposition., *Decision Analysis Unit, London,* Brunel University, Technical Report, 79-2/2.

Iacocca L., (1990). *An autobiography,* Bantam Books, Toronto.

Johnson E.,Schkade D.(1985) Bias in utility measurements:futher evidence and expectations.,*Management Science,*v.36,N 4,pp.406-424.

Jungermann H., (1983). The two camps on rationality, In: R. Scholz (*Ed.*), *Decision making under uncertainty,* Amsterdam, North -Holland.

Kahneman D., Slovic P., Tversky A. (*Eds.*), (1982). *Judgement under uncertainty: Heuristics and Biases.,* Cambridge University Press, Cambridge.

Kahneman D., Tversky A., (1979). Prospect Theory: An Analysis of Decisions under Risk., *Econometrica,* N 47, p. 263-291.

Keeney R. L., (1974). Multiplicative utility functions, *Operation Research,* 22, 22-34.

Keeney R. L., (1980). *Siting energy facilities,* N.Y.: Academic Press.

Keeney R. L., (1992). *Value-focused thinking,* Harvard Univ. Press.

Keeney R. L., Raiffa H., (1976). *Decisions with multiple objectives: Preferences and Value tradeoffs,* New York, Wiley.

Kelly G. A., (1955). *The psychology of personal constructs.* New York, Norton.

Kendall M., (1969). *Rank correlations,* New York: Academic press.

Kihlstrom J.(1987) The Cognitive Unconscious.,Science,v.237,pp.1445-1451.

Klatzky R. L., (1975). *Human Memory, Structures and Processes.*, W. Freeman and Co., San Francisco.

Larichev O. I., (1979). *Science and Art of Decision Making*, Nauka Publisher. (in Russian)

Larichev O. I., (1982). *A method for evaluating R&D proposals in large research organizations*, Collaborative paper CP-82-75, Laxenburg, Austria: IIASA.

Larichev O. I., (1984). Psychological Validation of Decision Methods. *Journal of Applied Systems Analysis*, 11, 37-46.

Larichev O. I., (1987). *Objective models and Subjective Decisions*. Moscow: Nauka (in Russian)

Larichev O. I., (1992). Cognitive validity in Design of Decision-Aiding Techniques, *Journal of MultiCriteria Decision Analysis*, v. 1, N 3, pp. 127-138.

Larichev O. I., Andreyeva E. N. ,Sternin M. Y., (1996). System for Preparing Management Decisions: a gas pipeline siting case study, In: P. Humpreys, L. Bannon, A. McCosh, P. Migliarese, J-P. Pomerol (*Eds.*), Implementing Systems for Supporting Management Decisions, London, Chapman & Hall.

Larichev O. I., Boichenko V. S., Moshkovich H. M., Sheptalova L. P., (1980). Modelling multiattribute information processing strategies in a binary decision task. *Organizational Behavior and Human Processes*, 26, 278-291.

Larichev O. I., Brown R., Andreeva L., and Flanders N., (1995a). Categorical Decision Analysis for Environmental Management: a Siberian Gas Distribution Case. In: Caverni J. P., Bar-Hillel M., Barron F. N. and Jungermann H. (*Eds.*), *Contributions to Decision Research*. North-Holland, Amsterdam, 255-279.

Larichev O. I., Furems E. M., (1987). Multicriteria Packing Problems., In: Y. Sawaragi, K. Inoue, H. Nakayama (*Eds.*), *Toward Interactive and Intelligent Decision Suppport Systems*, Springer Verlag, Berlin.

Larichev O. I., Mechitov A. I., Moshkovich H. M., and Furems E. M., (1989). *Expert Knowledge Elicitation.* Moscow: Nauka press. (in Russian)

Larichev O. I., Moshkovich H. M., (1980). On possibilities of eliciting consistent estimates of multidimentional alternatives from people. In: Emelyanov S. V. (*Ed.*), *Descriptive research into decision maing procedures under multiple criteria.* Moscow: VNIISI Press, 60-67. (in Russian)

Larichev O. I., Moshkovich H. M., (1986). Direct Classification Problems in Decision Making, *Doklady Akademii Nauk*, 287, 6, 567-570.

Larichev O. I., Moshkovich H. M., (1987). Direct Classification Method and Problems of Eliciting Reliable Expert Information, *Izvestia Akademii nauk SSSR. Technicheskaya kibernetika*, 1, 151-161 (in Russian)

Larichev O. I., Moshkovich H. M., (1988). Limits to decision making ability in direct multiattribute alternative evaluation., *Organizational Behaviour and Human Decision Processes*, N 42, pp. 217-233.

Larichev O. I., Moshkovich H. M., (1990). Decision Support System CLASS for R&D Planning. *Proceedings of First International Conference on Expert Planning Systems*, Brighton, Conf. Public. N 322, The Institue of Electrical Engineers, 227-232.

Larichev O. I., Moshkovich H. M., (1991). *ZAPROS: a method and a system for ordering of multiattribute alternatives on the basis of decision-maker's preferences.* Preprint, Moscow: VNIISI press. (in English).

Larichev O. I., Moshkovich H. M., (1992). Decision Support System ORCLASS. *Proceedings of the Tenth International Conference on Multiple Criteria Decision Making*, Taiwan, 1, 341- 350.

Larichev O. I., Moshkovich H. M., (1994). An Approach to Ordinal Classification Problems *International Transactions on Operational Research*, v. 1, N 3, pp. 375-385.

Larichev O. I., Moshkovich H. M., (1995). ZAPROS-LM-a method and system for rank-ordering of multiattribute alternatives., *European Journal of Operations Research*, v. 82, pp. 503-521.

Larichev O. I., Moshkovich H. M., (1996). *Qualitative Methods of Decision Making*, Moscow, Physmathlit Pulisher,(in Russian).

Larichev O. I., Moshkovich H. M., Drozdova I. V., Starinsky V. V., Kovalev B. N., (1990). Complex evaluation of R&D projects in health service and medicine, *Soviet Medicine*, 1, 32-37. (in Russian)

Larichev O. I., Moshkovich H. M., Furems E. M, (1986). Decision Support System "CLASS". In: B.Brehmer, H.Jungerman, P.Lourens and G.Seton (Eds.), *New directions on research in decision making*. Amsterdam: North Holland.

Larichev O. I., Moshkovich H. M., Furems E. M., Mechitov A. I., Morgoev V. K., (1991). *Knowledge Acquisition for the Construction of Full and Contradiction Free Knowledge Bases*. Groningen, the Netherlands: iec ProGAMMA.

Larichev O. I., Moshkovich H. M., Mechitov A. I., Olson D. L., (1993). Qualitative approaches to rank-ordering multiattribute alternatives, *Journal of Multi-Criteria Decision Analysis, 2*, 5-26.

Larichev O. I., Moshkovich H. M., Rebrik S. B., (1988). Systematic research into human behaviour in multiattribute object classification problems, *Acta Psychologica*, N 68, pp. 171-182.

Larichev O. I., Moshkovich H. L., Mechitov A. I., Olson D. L. (1993). Experiments comparing qualitative approaches to rank ordering of multiattribute alternatives, Journal of Multi-Criteria Decisioon Analysis,v. 2, N 1, pp. 5-26.

Larichev O. I., Olson D. L., Moshkovich H. M., Mechitov A. I., (1995b). Numerical vs. Cardinal Measurements in Multiattribute Decision Making: How Exact Is Exact Enough? *Organizational Behavior and Human Decision Processes*, 64, 1, 9-21.

Larichev O. I., Pavlova L. I., Osipova E. A., (1992). Multicrteria Problelms with Designed Decision Alternatives, In: A. Goicoechea , L. Duckstein, S. Zionts (*Eds.*), *Multiple Criteria Decision Making*, Springer Verlag, Berlin.

Larichev O. I., Petrovsky A. B., (1987). Decision Support Systems: the state-of-art., *Achivements in Science and Technology, Technical Cybernatics*, VINITI Publ., v. 21, pp. 131-165. (in Russian)

Larichev O. I., Polyakov O. A., Nikiforov A. D., (1987). Multicriterion linear Programming problems (Analytical Survey), *Journal of economical psychology*, N 8, pp. 387-407.

Larichev O. I., Zuev Yu. A., and Gnedenko L. S., (1974). Method for classification of applied R&D projects. In: S. V. Emelyanov (*Ed.*), *Perspective apllied R&D planning*. Moscow: Nauka press, 28-57. (in Russian)

Larichev O. I., Zuev Yu. A., Gnedenko L. S., (1978). Method ZAPROS (CLosed Procedures near Reference Situations) for the anlysis of variants of complex decisions, In: S. V. Emelyanov (*Ed.*), *Multicriteria Choice for the Solution of Ill-structured Problems,* Moscow: VNIISI Proceedings, N 5 (in Russian)

Lewicki P., Hill T., Czyzewska M.(1992) Nonconscious Acquisitioon of Information.,*American Psychologist,*June,pp.796-801.

Lindblom C. H., (1959). The science of mudding through, *Public Administration Review*, v. 19, v. 155-169.

Litvak B. G., (1982). *Expert information: methods for data collection and analysis*. Moscow: Radio and Svyaz. (in Russian)

Lootsma F. A., (1990). The French and the American school in multi-criteria decision analysis, Report 90-21, *Faculty of Technical Mathematics and Informatics, Delft University of Technology*.

Lotov A. V., Bushenkov V. A., Chernykh O. L., (1997). Multicriteria DSS for River Water-Quality Planning, *Microcomputers in Civil Engineering,* 12, pp.57-67.

Luce R. D., Raiffa H., (1957). *Games and Decisions : introduction and critical survey.,* New York, Wiley.

McCrimman K.R.,Wehrung D.A.(1975) Trade-off analysis:indifference and preferred proportions approaches.,In: D.Bell,R.Keeney,H.Raifa (Eds.) *Conflicting Objectives in Decisions,* New York,Wiley,pp.123-147.

Miller G. A., (1956). The magical number seven, plus or minus two :Some limits on our capacity for processing information., *Psychological Review*, 63, pp. 81-97.

Mirkin B. G., (1974). *Problems of group choice,* Moscow: Nauka press (in Russian)

Montgolfier de J., Bertier P., (1978). *Approche Multicritere des Problems de Decision*. Paris: Edicions Hommes et Techniques. (in French)

Montgomery H., Svenson O., (1989). A think-aloud study of dominance structuring in decision processes, In : H. Montgomery, O. Svenson (*Eds.*), *Process and Structure on Human Decision Making*, J. Wiley and Sons, Chichester, pp. 135-150.

Moshkovich H. M., (1985). A rational procedure of decision maker interview for assigning alternatives to decision classes. In: O. I. Larichev (*Ed.*), *Procedure and methods of decision making in management systems.* ,Moscow: VNIISI Press, 92-99.(in Russian)

Moshkovich H. M., (1988). Interactive system ZAPROS (for ordering of multiattribute alternatives on the basis of decision maker's preferences). In: Moscow: *VNIISI press,* 1, 13-21 (in Russian)

Nikiforov A.D.,Rebrik S.B.,Sheptalova L.P.,(1984) Experimental study of stability of preferences in some decision making tasks, In: S.V.Emeljanov, O.I.Larichev (Eds.)

Procedures for evaluation of multicriteria objects., VNIISI Proceedings,N 9,pp.69-80.(in Russian).

Novick D. (*Ed.*), (1965). *Program Budgeting*, Harward University Press.

Olson D. L., Courtney J. F., (1992). *Decision Support Models and Expert Systems*, Macmillan Publishing Co., New York.

Ore O., (1962). *Theory of graphs*, Rhode Island: American Mathematical Society.

Oseredko Y. S., Larichev O. I., and Mechitov A. I., (1982). Main Gas Pipeline Route Selection Problem. In: H. Kunreuther and E. Ley (*Eds.*) *Risk Analysis Controversy: An Institutional Perspective*, Berlin, Springer-Verlag, 91-101.

Osherenko, G. (1995). Indigeous political and property rights and economic/environment reform in northwest Siberia. *Post-Soviet Geography*, 36, 4, 225-237.

Ostanello A., (1990). Action evaluation and action structuring., In: C. Bana e Costa (*Ed.*) *Reading in Multiple Criteria Decision Aid*, Springer Verlag, Berlin , pp36-57.

Ozernoy V. M., Gaft M. G., (1978). Multicriteria problems of decision making. In: *Decision making problems*. Moscow: Mashinostroenie press, 14-47. (in Russian)

Payne J. W., Bettman J. R., Coupey E., Johnson E. J., (1993). A constructive process view of decision making :multiple strategies in judgment and choice, In: O. Huber, J. Mumpower, J van der Pligt, P. Koele (*Eds.*), *Current Themes in Psycological Decision Research*, North Holland, Amsterdam, pp. 107-142.

Quade E. S. (*Ed.*), (1963). *Analysis for Military Decisions*, RAND McNally and Co., Chicago.

Raiffa H., (1968). *Decision analysis. Readings*, MA, Addison -Wesley.

Rajkovic V.(1985) *Computer Based DSS for Nursery Schools (in Slovene).*, Internal report,IJSs DP-3920,Ljubljana,University of Ljubljana.

Rittel H. W., Webber M. M., (1973). Dilemmas in a General Theory of Planning., *Policy Sciences*, v. 4, pp155-169.

Rivett P., Ackoff R. (1963). A manager's guide to operational research., London, J.Wiley & Sons.

Roy B., (1985). *Methodologie Multicritere d'Aide a la Decision*, Economica, Paris (in French).

Roy B., (1996). *Multicriteria Methodology for Decision Aiding*, Kluwer Academic Publisher, Dordrecht.

Russo J. E., Rosen L. D., (1975). An eye fixation analysis of multiattribute choice, *Memory and cognition.*, N 3, P. 267-276.

Saaty T. L., (1980). *The Analytic Hierarchy Process*, New York, McGraw-Hill.

Simon H. A., (1960). *The New Science of Management Decision*, Harper and Row Publ.

Simon H. A., (1974). How big is a chunk?, *Science*, N183, p. 482-488.

Simon H. A., (1981). *The Science of Artificial.* London: MIT.

Simon H. A., (1991). Information -processing models of cognition, *Journal of Amer. Soc. Information Science*, Sept., pp. 364-377.

Stadler W., (1988). Fundamentals of Multicritera Optimization., In : W. Stadler (*Ed.*) *Multicrteria Optimization in Engineering and in the Sciences.*, Plenum Press, New York.

Statnikov R. B., Matusov J. B., (1995). *Multicriteria optimization aand engineering*, Chapman and Hall, New York..

Steuer R. E., (1986). *Multiple Criteria Optimization: Theory, Computation, and Application*, J. Wiley and Sons, New York.

Teigen K., (1988). The language of uncertainty, *Acta Psychologica*, 68, 27-38.

Tversky A., (1969). Intransitivity of Preferences, *Psychological Review*, 76, pp. 31-48.

Tversky A., Sattach S., Slovic P., (1988). Contingent weighting in judgment and choice, *Psychological Review*, 95, pp. 371-384.

Vari A., Vecsenyi J., (1994). Selecting Decision Suppot Methods in Organizations., *Journal of Applied Systems Analysis*, v. 11, p. 23-36.

Ventzel E. C., (1972). *Operations research*, Moscow, Soviet Radio Publisher. (in Russian)

von Neumann J., Morgenstern O., (1947). *Theory of games and economic behavior.* Priceton, NJ, Princeton University Press.

von Winterfeldt D., Edwards W., (1986). *Decision Analysis and Behavioral Research*, Cambridge University Press.

von Winterfeldt D., Fischer G.W.(1975) Multiattribute utility theory: Models and assessment procedures, In : D. Wendt, C. Vlek *(Eds.) Utilitty,probability aand human decision making,*Dordrecht,Reidel.

Wagner H. M., (1969). *Principles of Operations Research*, Prentice-Hall, Inc., Englewood Cliffs, New Jersey.

Wallsten T., Budescu D., Zwick R., (1993). Comparing the calibration and coherence numerical and verbal probability judgments, *Management Science*, 39, pp. 176-190.

Weber M.,Eisenfuhr F.,von Winterfeldt D.(1988)The effects of splitting attributes on weights in multiattribute utility measurement, *Management Science*, v.34.N4,pp.431-445.

Wilson R., (1972). *Introduction to Graph Theory*, Edinburgh: Oliver and Boyd press.

Wu G., (1993). Editing and Propect Theory: Ordinal Independence and Outcome Cancellation, *Working Paper of Harvard Business School*, 94-001.

Yu W., (1992). ELECTRE TRI, aspects methodologiques et manuel d'utilisation. In: *Document du LAMSADE* , 74, Paris, Universite de Paris-Dauphine.(in French)

Zuev Yu. A., Larichev O. I., Filippov V. A., Chuev Yu. V., (1979). Problems of evaluation of R&D projects, *Vestnik Academii nauk SSSR*, 8, 29-39 (in Russian)

INDEX

THEORY AND DECISION LIBRARY

SERIES C: GAME THEORY, MATHEMATICAL PROGRAMMING AND OPERATIONS RESEARCH

Editor: S.H. Tijs, *University of Tilburg, The Netherlands*

17. O.I. Larichev and H.M. Moshkovich: *Verbal Decision Analysis for Unstructured Problems*. 1997 ISBN 0-7923-4578-9